BITING MIDGES IN THE CRETACEOUS AMBER OF NORTH AMERICA

(Diptera: Ceratopogonidae)

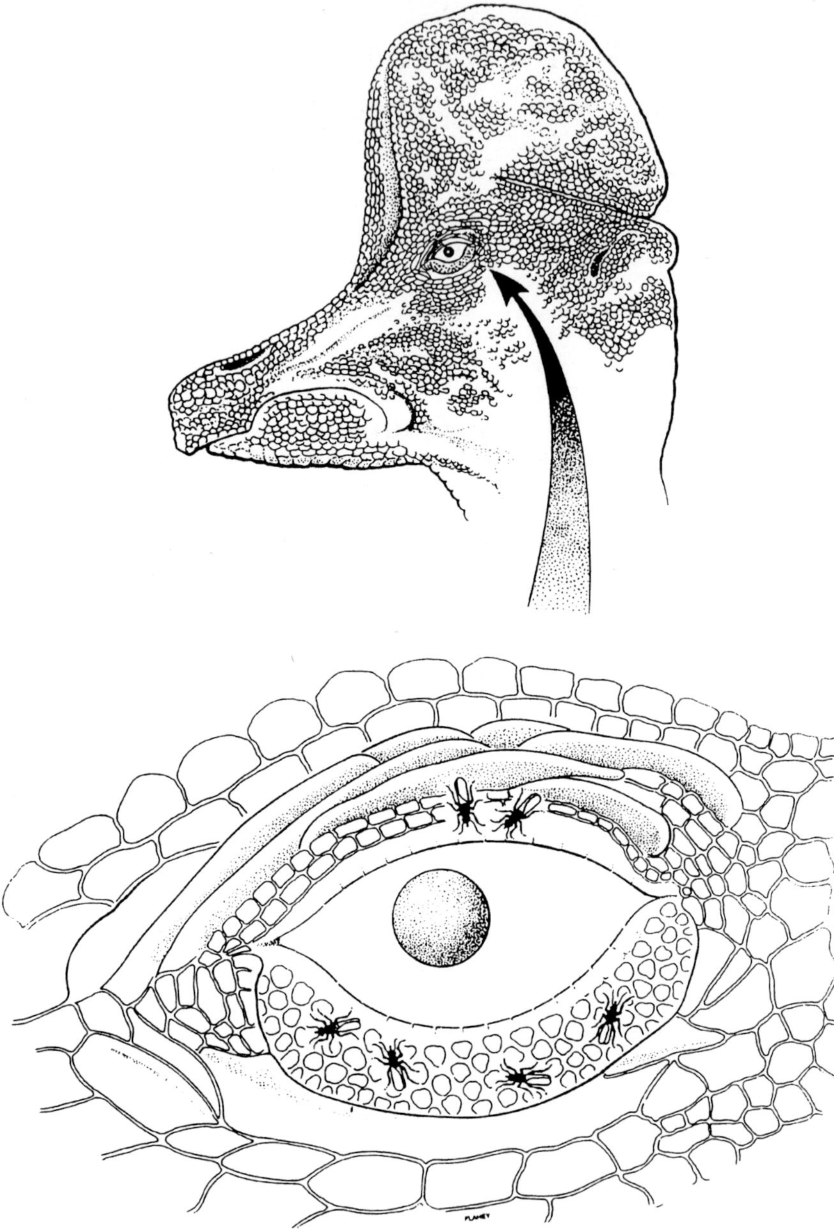

Female *Culicoides canadensis* sucking blood from the eyelids of a *Corythosaurus* in southern Alberta, 72-84 million years ago.

BITING MIDGES IN THE CRETACEOUS AMBER OF NORTH AMERICA (Diptera: Ceratopogonidae)

BY

A. BORKENT

BACKHUYS PUBLISHERS
LEIDEN
1995

Cover illustration: Female adult of *Culicoides variipennis* (Coquillett).

Reproduced from Volume 1 of the Manual of Nearctic Diptera (1981, Research Branch of Agriculture Canada Monograph No. 27) with the permission of the Minister of Supply and Services Canada, 1994.

CIP-data Koninklijke Bibliotheek, Den Haag

Borkent, A.

Biting midges in the Cretaceous amber of North America
(Diptera: Ceratopogonidae) / by A. Borkent. - Leiden :
Backhuys Publishers. - Ill.
With index, ref.
ISBN 90-73348-40-4 bound
NUGI 823/821
Subject headings: midges ; North America / amber ;
North America / palaeontology.

Printed in The Netherlands.

Dedicated to the memory of

Willis Wagner Wirth
(1916-1994)

in recognition of his outstanding contributions to the systematics of biting midges and his unfailing support of the research of all his colleagues.

TABLE OF CONTENTS

ABSTRACT

Borkent, A.
The Ceratopogonidae found in Upper Cretaceous amber from Canada and New Jersey, U.S.A. are described, keyed, and analyzed. Eighteen species are recognized in Canadian amber, most of these as both males and females (and one with eggs) and one as a previously described species in New Jersey amber. Some biological information is available for some of these species. To interpret phylogenetic relationships of all Cretaceous fossils, a cladistic analysis of extant Ceratopogonidae is provided. Some tribes of Ceratopogoninae form a grade, indicating that the present higher classification is inadequate. Canadian amber ceratopogonids are early lineages within the family, as are Cretaceous fossils from other amber deposits. Comparison with Tertiary amber Ceratopogonidae (Baltic, Saxonian, and Dominican Republic) shows that much of the extant generic diversification of the family occurred between the Late Cretaceous and Eocene. The family itself likely originated and initially diversified in the Jurassic.

Comparisons of Canadian amber fossils with other Cretaceous ceratopogonids indicates relationships with those described from the Taimyr Peninsula, Russia. The limited fauna from French amber, somewhat older than Canadian and Taimyr amber, includes only the very earliest lineages of the family.

Examination of the character states of all biting Nematocera indicates that blood-sucking is plesiotypic to the Ceratopogonidae and that all early lineages fed on vertebrate blood. Feeding on the blood of other invertebrates is likely to have arisen at least three times within the family. Comparison of details of the mouthparts of the fossils with extant *Culicoides* Latreille indicates that two and possibly three fossil *Culicoides* species fed on large dinosaurs. Other features of the biology of the fossil species are also discussed. The species formed an important part of the palaeoenvironment in southern Alberta during the Late Cretaceous. In general, their surroundings were a subtropical to tropical environment with seasonal variations in precipitation.

Canadian amber includes the following five new genera, *Adelohelea, Heleageron, Minyohelea, Palaeobrachypogon* and *Peronehelea* and 11 new species, *Leptoconops primaevus, Adelohelea glabra, Culicoides agamus, C. annosus, C. bullus, C. obuncus, Heleageron arenatus, Minyohelea pumilis, Palaeobrachypogon remmi, P. vetus* and *Peronehelea chrimikalydia*.

Atriculicoides cretea (Boesel) is a new junior synonym of *Atriculicoides globosus* (Boesel). The following are new combinations: *Atriculicoides cretea* (Boesel), *Atriculicoides globosus* (Boesel), *Culicoides canadensis* (Boesel), *Culicoides tyrrelli* (Boesel), *Palaeobrachypogon aquilonius* (Boesel), *Palaeobrachypogon frigidus* (Remm) and *Palaeobrachypogon macronyx* (Remm).

As a new combination, *Culicoides canadensis* (Boesel) is a senior homonym of *Culicoides canadensis* Wirth & Blanton (1969:211) which is given the new name *Culicoides paraimpunctatus*.

INTRODUCTION

The insects found in amber provide a striking and remarkable vista of life in the past. Unlike the remnants of bones and teeth studied by vertebrate palaeontologists, insects in amber are often in pristine condition, looking for all intents and purposes like freshly captured material. These fossils are therefore a wonderful source of information on the evolution of taxa, provide insights on past environments and furnish evidence of former ecological associations and adaptations.

The best known amber faunas are those of the Tertiary. Baltic amber, of Eocene-Lower Oligocene age, is particularly well known, with hundreds of thousands of insects present and many of these described (Larsson, 1978; Krzeminska and Krzeminski, 1992). Miocene-Oligocene amber deposits from Chiapas, Mexico (Hurd, et al., 1962; Poinar, 1992:37-46, 279-283), Dominican Republic (Baroni-Urbani and Saunders, 1982; Poinar, 1992:29-37, 284-288) and Bitterfeld, Germany (Saxonian amber) (Schumann and Wendt, 1989) are other major Tertiary records of insect evolution. These taxa are either generically indistinguishable from our modern fauna or include a few genera unknown in the extant fauna. At least some of the 'unknown' genera may merely be autapomorphic species which, once critical cladistic analysis has been applied, may prove to belong within extant genera.

Canadian amber is of Upper Cretaceous age (71-84 million years old), and provides a splendid opportunity to examine insect diversification at a significantly earlier stage than previously described Tertiary amber faunas. Like the Upper Cretaceous amber from France (Schlüter, 1978), the Taimyr Peninsula in Russia (Zherikhin and Sukatsheva, 1973), and from New Jersey (Grimaldi, et al., 1989), included insects are often of early lineages within extant families or are placed in entirely new families.

Most families of insects found in Canadian amber are represented by only a few specimens (Beardsley, 1969; Botosaneanu and Wichard, 1983; Brown and Pike, 1990; Cooper, 1964; Gagné, 1977; Hamilton, 1971; Klimaszewski and Kevan, 1986; Krzeminski and Teskey, 1987; MacKay, 1970; McAlpine, 1973; McAlpine and Martin, 1966; Peterson, 1975; Richards, 1966; Schawaller, 1991; Teskey, 1971; Wilson, 1985; Yoshimoto,

1975). The most common group of insects is the aphids (Heie and Pike, 1992; McAlpine and Martin, 1969:836-837; Pike, 1994). The second most abundant family in Canadian amber is the Ceratopogonidae and this group has not been studied since the description of six species (based on seven specimens) by Boesel in 1937. The presence of 261 specimens available for the current study provided an unprecedented opportunity to acquire insights into the evolutionary and ecological diversification of this group. True to initial expectations, there were a number of delightful surprises, including initial evidence from mouthpart morphology that some female Cretaceous Ceratopogonidae took blood meals from dinosaurs.

This study describes 18 species of Ceratopogonidae from Canadian amber (Table 1), generally as both sexes (n=13), and one species from New Jersey amber, places these within the context of a cladistic analysis of the family, interprets the implications of these relationships for our understanding of the diversification of the family, and examines the evidence interpreting the evolution of biting habits. The habitats and habits of individual species are analyzed and these are placed in a broader context of our understanding of the general environment present in western Canada during the Late Cretaceous.

Cretaceous ambers including Ceratopogonidae are known from seven areas (Table 2): France (Szadziewski and Schlüter, 1992); Austria (Schlee, 1984); Lebanon (Grimaldi, 1993a; Poinar et al., 1994; Schlee and Dietrich, 1970); the Taimyr Peninsula, Russia (Remm, 1976); Alaska (Langenheim, et al., 1960, as Heleidae); Grassy Lake, Alberta and Cedar Lake, Manitoba in Canada; and New Jersey, U.S.A. (Grogan and Szadziewski, 1988). Species have been described from all of these areas except from the deposits in Lebanon, Austria and Alaska (which were merely identified to the family level). This investigation allowed for a comparison of these Cretaceous faunas with each other and with younger Tertiary fossils, permitting some generalizations to be made about diversification within the family.

The characteristics and features of New Jersey amber were thoroughly discussed by Grimaldi et al. (1989) and are not repeated here.

ACKNOWLEDGEMENTS

Foremost, my sincere and heartfelt thanks goes to my wife Annette for her continuing recognition of the value of systematic studies such as this. She supported my work with faith, encouragement, love, and (not least) finances. Without her, these Cretaceous fossils would not have been described and analyzed in this book!

I express my appreciation to the curators who lent me the material upon which this study is based: Dr. F.M. Carpenter (MCZC), Dr. J.M.

Cumming (CNCI), Mr. E.M. Pike (TMPD), Mr. R.J. Pupedis (PMNH), and Drs. G.B. Wiggins and D.C. Darling (ROME).

Several individuals kindly identified inclusions present in amber pieces otherwise containing Ceratopogonidae: Dr. R. Stockey (University of Alberta), *Glyptostrobus*; Dr. L. Masner (CNCI), Serphitidae; Dr. M.J. Sharkey (CNCI), Ichneumonoidea; Dr. I.M. Smith (CNCI), Acari; and Dr. R.G. Foottit (one specimen of *Mesozoicaphis*; others identified by author).

The masterful frontispiece, showing how life may have been shared by a dinosaur and a biting midge in the distant past, was drawn by Mr. Barry Flahey (CNCI) and I thank him for this assistance.

I also appreciate the help I received from Dr. Neal L. Evenhuis in locating some obscure literature and from Mr. Jim Troubridge who provided labels for specimens.

Dr. Richard C. Fox kindly commented on the characteristics of mammals and environmental conditions in southern Alberta during the Late Cretaceous. Dr. D.A. Russell provided information on the characteristics of dinosaur skin and potential sites where biting midges might bite ceratopsians. Dr. Beverly K. Mitchell gave advice concerning details of sensillae structure and function. Mr. A.J. Ross kindly wrote with news that the lost fossil *Simulidium priscum* has been rediscovered in the BMNH.

Dr. D.A. Duckhouse furnished advice and copies of papers dealing with the mouthparts and feeding behaviour of Psychodidae.

Mrs. Pam Giacomello assisted with the printing and copying of the manuscript and her help is gratefully acknowledged.

I also thank my son Michael for assistance with compiling data on numbers of specimens and their statistics.

Last but certainly not least I express my appreciation to Drs. D.A. Grimaldi, W.L. Grogan and the late W.W. Wirth for providing much valued critiques of an earlier draft of this work.

SOURCE OF MATERIAL, STRATIGRAPHY, AND AGE OF CANADIAN AMBER

Canadian Cretaceous amber containing insect inclusions has been found at two localities: 5 km south of Grassy Lake, Alberta and Cedar Lake, Manitoba (Fig. 20B).

Grassy Lake amber is found amongst coal mine tailings and likely formed in situ (Heie and Pike, 1992). The lignite deposits are from the Foremost Formation of the Belly River Group and radiometric dating (potassium-argon) of the Bearpaw shales immediately above the Foremost deposits at Medicine Hat (67 km east of Grassy Lake) are dated at 72-73 million years (Folinsbee, et al., 1964). The Foremost Formation belongs within the Campanian Stage which is dated by Obradovich (1993) as 71.3 ± 0.5 - 83.5 ± 0.5 million years old. Grassy Lake amber is therefore 72-84 million years old.

McAlpine and Martin (1969) provided a detailed description of the Cedar Lake amber, its origins, characteristics, inclusions (at a family level) and the previous expeditions made to collect specimens at the site. All the Cedar Lake specimens reported in this study originated from a single beach, although small amounts of amber are scattered throughout the general vicinity (Klotz, 1892). Poinar (1992:51) provided a detailed map of the area at the west end of Cedar Lake but mistakenly identified the amber beach as too far south; the beach (now flooded by a dam) was actually on the north side of Oleson's Point, directly opposite the mouth of the Saskatchewan River.

Cedar Lake amber was obviously secondarily deposited. The Saskatchewan River, which deposited the amber, originates in the Rocky Mountains of western Alberta, flows through Alberta, with one of its two major tributaries, the South Saskatchewan River, flowing within 15 km of the Grassy Lake site, continues eastward through Upper Cretaceous deposits and, as its first major settling basin, forms Cedar Lake in Manitoba (Fig. 20B). The final 240 kilometers of its flow before entering Cedar Lake is through Silurian deposits but these are clearly too old to be the source of the amber. Small amounts of amber have also been found along the length of the Saskatchewan River which are indistinguishable from Cedar Lake amber (McAlpine and Martin, 1969:827), confirming the likelihood that the amber was transported downstream from Alberta.

There is limited evidence that the Cedar Lake amber also originated in Campanian deposits in Alberta and, as such, that the amber is about 72-84 million years old (i.e. the same as material from Grassy Lake). Based on infrared absorption spectra, Langenheim and Beck (1968) indicated that 10 out of 11 pieces of Cedar Lake amber were of one type, corresponding to that from New Jersey amber, and one piece was somewhat similar to Baltic amber. The second type may indicate either another original deposit or a different source tree. Pike (1991) confirmed that Cedar Lake amber was of two types and that one of these (presumably the most common type) was similar to the amber from 5 km south of Grassy Lake, Alberta. Details of the infrared spectrum of the second type have not been published, preventing any further comparisons to be made at this time. Nevertheless, this similarity suggests that Cedar Lake amber is of comparable age to that from Grassy Lake.

Furthermore, evidence from a shared insect fauna further supports the hypothesis that the Grassy Lake and Cedar Lake amber are of the same age. Yoshimoto (1975) reported that the chalcidoid fauna from 5 km south of Grassy Lake (given as "near Medicine Hat") and Cedar Lake was similar. Of the 12 species that he recognized (based on 54 specimens), seven were present in both ambers and five were found in Cedar Lake amber only. However, the five species restricted to Cedar Lake amber were represented by only 1-3 specimens each, while the species present in both ambers numbered 5-12 specimens each. The current study of Ceratopogonidae similarly indicates that nearly all species are present in both ambers. Out of a total of 18 species, only three are restricted to either Cedar Lake or Grassy Lake amber and these are represented by only a few specimens: *Culicoides bullus* n.sp. (n= 3), *C. agamus* n.sp. (n= 3) and *Minyohelea pumilis* n.sp. (n= 7 but 6 of these were in one piece of amber). This shared fauna indicates that Cedar Lake amber is of the same age as that from 5 km south of Grassy Lake. Other studies are based on so few specimens, that further comparisons are not yet possible (Beardsley, 1969; Botosaneanu and Wichard, 1983; Brown and Pike, 1990; Cooper, 1964; Gagné, 1977; Hamilton, 1971; Heie and Pike, 1992; Klimaszewski and Kevan, 1986; Krzeminski and Teskey, 1987; MacKay, 1970; McAlpine, 1973; McAlpine and Martin, 1966; Peterson, 1975; Richards, 1966; Schawaller, 1991; Teskey, 1971; Wilson, 1985; Yoshimoto, 1975).

In this regard, it is unfortunate that the aphid fauna in Canadian amber has not been more comprehensively studied as these are the most common group preserved (about 40% of all arthropod inclusions in Canadian amber; McAlpine, pers. comm.). Specimens are very common in the amber (Pike, 1994); in the CNCI alone there are over 800 specimens, most in good to excellent condition (pers. obs., McAlpine, pers. comm.).

BOTANICAL SOURCE OF CANADIAN AMBER

Because of comparisons made between different ambers and the detailed analysis of New Jersey amber by Grimaldi et al. (1989), the source of the resin which produced Canadian amber can be identified with reasonable certainty.

McAlpine and Martin (1969) described the physical features of Cedar Lake amber in some detail. Langenheim (1969, 1990) and Langenheim and Beck (1968), based on infrared spectra analysis, reported that most Cedar Lake (10 out of 11 pieces) and New Jersey amber were similar to each other and to extant *Agathis* and hence that both likely originated from a Cretaceous Araucariaceae. Broughton (1974) confirmed that the infrared spectra of amber from Cedar Lake and New Jersey were similar to that of *Agathis*. Poinar and Haverkamp (1985), using pyrolysis mass spectrometry on two pieces of Cedar Lake amber, also showed that these were very similar to the resin-copal produced by *Agathis* and that the amber was also very similar to that from Alaska; their study did not include material from New Jersey. Shedrinsky et al. (1991) illustrated the pyrolysis gas chromatogram of Cedar Lake amber as well as that from other sources. Amber from Cedar Lake was examined by Lambert et al. (1990) using nuclear magnetic resonance spectroscopy and showed significant similarities to amber from Alaska and New Jersey, indicating that they all came from a common botanical source. Further analysis (using interrupted decoupling) suggested even greater similarities between Cedar Lake and New Jersey amber. Finally, Grimaldi et al. (1989), using a variety of analytical tools, have identified New Jersey amber as araucariaceous in origin. This confirms that Cedar Lake amber also likely formed from Araucariaceae resin.

Although the specific characteristics or tree of origin of the Grassy Lake amber have not been previously reported, it seems likely that this amber, similar in appearance to that from Cedar Lake and sharing a common fauna (see above under "Source of material, stratigraphy, and age of Canadian amber"), also came from an Araucariaceae. Heie and Pike (1992) reported that Grassy Lake amber had one type of infrared absorption spectra, indicating a single plant source. Pike (1991) indicated that the infrared absorbtion spectra of Grassy Lake amber was similar to one of two types of amber identified from Cedar Lake (presumably the most common one). Neither Heie and Pike (1992) nor Pike (1994) provided the actual spectrum itself.

Araucariaceae macrofossils are known from the Jurassic of Utah (Stockey, pers. comm.), the Upper Cretaceous beds in Wyoming (Dorf, 1942) and Upper Cretaceous deposits on the Atlantic coastal plain of the U.S.A. (Penny, 1947). The family became extinct in the Northern Hemisphere during the early Tertiary (Langenheim, 1969). Stockey (1982) has further noted, however, that fossil *Agathis*, the identified source of Canadian amber, has never been otherwise identified in the northern hemisphere (northern fossils are of *Araucaria* or extinct genera). This may indicate that the amber source is incorrectly identified as being *Agathis*.

No comprehensive analysis of plant specimens in Canadian amber has been made, although such specimens are relatively common in the amber. Some botanical inclusions are noted in some species descriptions below; a possible seed was in one piece, trichomes were present with a number of ceratopogonids and one piece of amber contained a *Glyptostrobus* twig. Such twigs were also present in other pieces of Canadian amber which I examined but which did not include Ceratopogonidae. Considering that the single extant species of *Glyptostrobus* produces only moderate amounts of resin, it is unlikely to be the source of Canadian amber.

MATERIALS AND METHODS

Specimens examined. This study was based on the examination of 111 male and 143 female Ceratopogonidae in Upper Cretaceous Canadian amber, with 66 and 93 of these, respectively, identified to species (Table 1). I also studied the single type specimen of *Culicoides casei* Grogan and Szadziewski from New Jersey amber. I was unable to examine critically important Ceratopogonidae amber fossils described from the Upper Cretaceous of the Taimyr Peninsula by Remm (1976; see Table 18) and housed in the Paleontological Institute of the Russian Academy of Sciences (formerly the Soviet Academy of Sciences) in Moscow, Russia. Fortunately, these are being studied by R. Szadziewski and will certainly further our understanding of Cretaceous Ceratopogonidae. Nor was I able to examine the Ceratopogonidae from Cretaceous Alaskan amber noted by Langenheim et al. (1960) as these could not be located (not held by R.L. Langenheim (pers. comm.), nor present in the Museum of Paleontology, University of California, Berkeley, California: G. O. Poinar, pers. comm.).

The abbreviations used to represent various museums cited in this text are those provided by Arnett and Samuelson (1986) or are abbreviated according to their procedure. Those used in this study are as follows:

BMNH - Department of Entomology, Natural History Museum, London SW7 5BD, United Kingdom.

CNCI - Canadian National Collection of Insects, Centre for Land and Biological Resources Research, Agriculture Canada, K.W. Neatby Building, Ottawa, Ontario K1A 0C6, Canada.

MCZC - Entomology Department, Museum of Comparative Zoology, Harvard University, 26 Oxford St., Cambridge, Massachusetts 02138, USA.

OXAM - Hope Entomological Collections, University Museum, Parks Road, Oxford 0XI 3PW, United Kingdom.

PMNH - Peabody Museum of Natural History, Yale University, New Haven, Connecticut 06520, USA.

ROME - Department of Entomology, Royal Ontario Museum, Toronto, Ontario M5S 2C6, Canada.

TMPD - Tyrrell Museum of Palaeontology, Box 7500, Drumheller, Alberta T0J 0Y0, Canada.

Specimens of Ceratopogonidae in Canadian Cretaceous amber are housed in four institutions: CNCI, MCZC, ROME and TMPD. Each of these museums has a numbering system for their material and these are reported for each specimen below. Amber pieces containing arthropod inclusions housed in the CNCI are given CAS (Canadian Amber Specimen) numbers.

To understand the phylogenetic placement of fossil species, adults of most extant genera of Ceratopogonidae were examined.

Collection of material. All the specimens of Ceratopogonidae in Canadian Cretaceous amber were collected at either 5 km south of Grassy Lake, Alberta or Cedar Lake, Manitoba.

Amber pieces from 5 km south of Grassy Lake, Alberta are scattered in and on large piles of old coal mine tailings. As such, this material has been collected by walking on and around these tailings and hand picking or sifting material. Most specimens from 5 km south of Grassy Lake were collected by staff at the CNCI: J.F. McAlpine, H.J. Teskey, B.E. Cooper, and D. Daze. The few specimens studied from the TMPD were collected by their staff using hand picking and sifting (Pike, 1994).

The amber from Cedar Lake, Manitoba was collected by hand picking through beach debris on the north side of Oleson's Point or with the use of a sluice (F.M. Carpenter, pers. comm.) and was extremely abundant; Tyrrell (1892) estimated the presence of 1,457,280 pounds of amber. McAlpine and Martin (1969) provided a detailed description of the expeditions made to collect specimens at the site. Although the original location on the north side of Oleson's Point was tragically flooded by the erection of a dam at Grand Rapids, Manitoba in 1969, at least some material has since been retrieved from the lake. I examined a small collection made by H. Kutzner, Grand Rapids in 1988 and J.F. McAlpine found a very small amount of material (two ounces collected in about 14 man hours of work) at Oleson's Point (now Oleson's Island) in 1983 (McAlpine, pers. comm.).

Preparation, examination techniques and measurements. Nearly all the specimens examined in this study were previously identified as at least Diptera, most often as Ceratopogonidae, sometimes as Chironomidae and occasionally as another family of Nematocera.

The pieces of amber were initially either stored on cotton in small boxes (CNCI), in mineral oil in vials (ROME) or were mounted in Canada Balsam on microscope slides (MCZC, ROME, TMPD). Nearly all needed further treatment before detailed observations could be made. For those which were mounted on microscope slides, the amber pieces were first cleaned of surrounding Canada Balsam by soaking in xylene. Mineral oil was removed with soap and hot water. After cleaning, amber pieces were treated as follows. If the amber piece was fairly large, or contained other inclusions, the piece was cut with a Sears Craftsman variable speed rotary power tool. This relatively slender, hand held tool rotated a thin (0.6 mm), circular (24 mm in diameter) hard disc which provided an excellent means for cutting the amber into small pieces. In addition, the flat surface of these discs could be used to slowly grind a broader surface. This allowed the amber to be fashioned into appropriate shapes to allow for maximal observation of inclusions. Amber pieces were cut to allow the best possible views of wings, legs and, especially for males, of genitalia. After cutting or grinding, the pieces were further ground by rubbing these along the length of a 6 cm long metal, fine nail file held flat on a firm surface. Generally amber pieces were ground to about a 1-2 mm thickness and finally polished using tin oxide (polish grade) on a leather belt. Each piece was then washed with soapy warm water to remove the tin oxide (sometimes with a soft toothbrush), dried, and mounted on a microscope slide in Canada Balsam. Care was taken to ensure that the coverslip was perpendicular to the line of vision as this markedly improved resolution.

It was striking that Canadian amber appeared to be substantially harder, and therefore requiring more time to grind and polish, than Baltic, Dominican Republic and New Jersey amber (pers. obs.).

It was necessary to examine a few specimens from a number of angles and in such instances the amber was ground to several flat planes (depending on the orientation of those structures being studied), placed on a microscope slide (sometimes held in place by a small amount of plasticine), and a small amount of immersion oil with a piece of coverslip was placed over the area requiring examination. These specimens were later mounted permanently in Canada Balsam.

It is apparent that old pieces of Canadian amber left exposed to air or placed in mineral oil become increasingly fragile and develop numerous surface fractures. Material in the CNCI which was collected in the 1970's and continuously exposed to air was in good condition but amber collected in 1950 from Cedar Lake was highly fragile. However, material collected in the 1930's and stored in Canada Balsam on microscope slides in the MCZC was in excellent condition. Szadziewski (1988:14) noted the same deterioration of Baltic amber which was exposed to air but that amber preserved in Canada Balsam since 1859 (Thomas collection, Museum für Naturkunde, Berlin) was in excellent condition. Grimaldi (1993b) also discussed such deterioration and that Canada Balsam was a good preserving agent of amber.

It is clear, therefore, that all amber material should be permanently stored in Canada Balsam. If a different orientation is required in the future, the specimens can be easily soaked free in xylene and reoriented.

Specimens were examined with both a Wild M3 dissecting microscope (up to 90X magnification) and a Carl Zeiss Jenaval compound microscope. Because all specimens were in thin pieces of amber, these could be examined at 312X and in some cases at 500X. Individuals were measured and drawn with the compound scope. In many instances the use of additional incidental light improved viewing with the compound scope, especially when the specimens were slowly rotated on the stage. This approach highlighted features not otherwise visible.

Measurements were taken with structures in as perpendicular a position as possible. Clearly, few specimens were in such an ideal position and this is almost certainly reflected in a higher than normal variation (i.e. compared to extant specimens dissected and mounted on slides). Because of the often awkward position of specimens within the amber, the measurements of various structures were sometimes taken from different sides of the ceratopogonid. For example, palpal segment three may have been measured on the right palp and palpal segment four on the left. This, in some instances, may also increase the apparent variation reported for a species.

Mouthpart length was measured as the distance from the most ventromedial ommatidia to the tip of the labrum. The measurement was clearly a difficult one to take accurately and is likely affected by the state of preservation of the head capsule. Nevertheless, it gave an approximate estimate of the striking variation in mouthpart length of some Cretaceous Ceratopogonidae. Furthermore, previous authors working with extant *Culicoides* Latreille (e.g. Blanton and Wirth, 1979), have estimated the relative length of the proboscis by dividing the length of the labrum by the distance from the interocular seta to the torma. In this study, this head capsule measurement was nearly always impossible to take. Therefore the relative length of the proboscis was calculated by dividing the length of the labrum by the length of the fifth tarsomere of the foreleg (which was considered to be of relatively constant length in relation to overall body size. Antennal, palpus, and wing measurements were taken as described by Grogan and Wirth (1979a:2).

A major difficulty in providing accurate measurements of fossil ceratopogonids was the apparent modification by deterioration and/or compression of some specimens. The presence of obvious asymmetry of palpal shape, antennal flagellomere shape, the degree of contrast of wing veins and the presence of extraneous material on some setae (making them appear thick) in some individuals warned of the possibilities of misinterpreting some structural differences.

Terms for structures follow those used in the Manual of Nearctic Diptera (McAlpine, 1981; Downes and Wirth, 1981). The term "not visible" is used in the species descriptions to indicate that a given character could not be seen clearly enough to allow for a character state to be reported.

Species descriptions include a characterization of the degree of dorsomedial separation of the eyes on the head capsule as eyes either broadly separated, narrowly separated or broadly contiguous dorsomedially. Broadly separated refers to the state present in extant *Leptoconops* Skuse (Downes and Wirth, 1981: Fig. 4); narrowly separated is the state present in species where the eyes are separated by a distance of one to about four ommatidial diameters (present in most extant Ceratopogonidae); and broadly contiguous is the condition present in most extant Forcipomyiinae and *Dasyhelea* Kieffer where a number of ommatidia are abutting, or nearly abutting, dorsomedially.

Difficult to interpret character states. Some features were particularly difficult or awkward to observe accurately and the reader should be aware that description of these features here may be inaccurate or report incorrect variation within a given species.

Previous reports on the presence or absent of eye pubescence may depend to a certain degree on the state of the specimen being observed. As noted below under the 'Taxonomic Discussion' of *Atriculicoides* Remm, some specimens initially appeared to have pubescent eyes but with further cutting and polishing of the amber, were found to lack this feature. The presence of surface contamination and distortion by the amber may give the false impression that pubescence is present when it is actually missing. None of the specimens examined for this study had pubescence.

The presence of sensilla on the antennal flagellomeres was often difficult to observe. Because of their small size and the varying state of preservation of the fossils, the sensilla coeloconica could be seen on only a few specimens.

The presence of pruinosity on the scutum could be determined for only a very few specimens when the thorax of these specimens were viewed from a lateral perspective in very thin and/or clear pieces of amber.

The presence and size of the humeral pits was difficult or impossible to see accurately in the material studied here. It is clear that a number of artifacts (surface contamination, distortion by the amber, compression of the fossil) make it nearly impossible to be certain of this character state for most Canadian Cretaceous fossils.

The presence of inner teeth on the claws of species with small claws was often difficult to confirm. Accurate characterization of this feature requires that specimens be observed at a high magnification and that the claws be oriented in the proper position. The presence of an empodium was also often difficult to confirm. In some specimens, one or more legs would have a well developed empodium while it appeared to be absent from the other legs. This is almost surely an artifact of preservation and/or observation. Finally, characterization of the tips of male claws as either bifid or not was also very difficult. In certain fossils, some claws appeared bifid while others in the same individual appeared simple. These leg characters are sometimes difficult to appraise accurately even in extant slide material.

Previous authors have pointed out the significance of male cerci which are directed posteriorly (Szadziewski, 1988; Grogan and Szadziewski, 1988). A few specimens examined in this study indicated that this feature is intraspecifically variable (e.g. *Culicoides filipalpis* Remm). It is likely that posteriorly directed cerci are merely an artifact and are distended because of their preservation in amber (or perhaps from partial decay before final preservation) (Borkent and Grogan, in press).

Criteria for species recognition. Clearly, the recognition of most fossil species is dependent on an interpretation of morphological structures. Ceratopogonids have been found in copula in amber (Davis, 1989; Montgomery de Merette, 1984) but all the males and females in Cretaceous amber were isolated from one another. Character states which were recognized as useful in distinguishing modern taxa as well as those which, if a series of fossils were available, showed discontinuities of variation (i.e. evidence of genetic discontinuity), were used to recognize taxa in this study.

Because of substantial collections made in the past, many of the species were represented by a series of specimens. These were often in varying states of preservation. Specimens were either completely opaque (well preserved or partially collapsed), partially cleared (probably partially decomposed), or were modified by what appeared to be heat and/or compression. The latter often had finely fractured cuticle, were of a characteristic brown color, nearly always incomplete and at least somewhat distorted. Such variation in preservation sometimes allowed for structural details to be examined which would otherwise not have been possible. However, such varied preservation, as well as differences in orientation of specimens within the amber, provided real challenges for interpreting morphological structures and hence, species concepts.

In general, a broad species concept was applied to the material. It is highly likely that once further material becomes available, some of the species described here will actually represent two or more species. I have described striking morphological variation under the 'Taxonomic discussion' of the pertinent species.

Descriptive format. Described character states are those which differ between congeneric species, are significant characters states distinguishing species from other Cretaceous species, or were of phylogenetic importance. In those instances where fossils species are placed in extant genera (i.e. *Leptoconops, Culicoides*), generic character states are included in the species descriptions to confirm these states in the fossils. Similarly, character states are described fully for both males and females to detail the evidence indicating that males and females were correctly associated.

The presence of other inclusions was noted for each species. Some of these give further information on the type of habitat in which the amber originated and some associations suggest adult behaviours.

The collection site at 5 km south of Grassy Lake has been referred to in previous publications and on specimen labels as either "near Medicine Hat" or just "Grassy Lake" (McAlpine, pers. comm.) in an attempt to thwart illegal collecting. In this study, I report the location as recorded on the specimen label followed by the true collection site in square brackets.

The collection number of each amber piece was recorded for each species, according to the system used by the lending institution. For the CNCI these numbers were preceded by the letters CAS (= Canadian Amber Specimen), for the MCZC by MCZ and for the ROME by ROM. If an amber piece holding more than one specimen was subsequently divided into two and mounted on separate slides, the amber pieces retained their original number but were further identified with an 'a' and 'b'.

All phylogenetic relationships are discussed in the separate section "Phylogenetic Analysis".

KEYS TO ADULT CERATOPOGONIDAE IN CANADIAN CRETACEOUS AMBER

If an unidentified specimen is in an awkward position, is incomplete or is too opaque to observe some details, it may not be possible to use the keys below. In such instances the individual may yet be identified using character states which are unique or restricted to a few species and which are listed below, following the keys.

Many amber specimens are distorted, partially crushed or are otherwise modified and these may look strikingly different. Readers are encouraged to read the pertinent section in materials and methods ("Difficult to interpret character states") before using the keys below.

Males

The males of *Palaeobrachypogon aquilonius* (Boesel) and a second species of *Atriculicoides* (identified here as *Atriculicoides* sp.) are unknown.

1. Wing with macrotrichia present over most of wing surface (Figs 3A,7A) or with macrotrichia along all major veins and some on membrane in at least cells m_2, cua_1, cua_2 (Figs 9A,10A); always some macrotrichia present on membrane on basal half of wing 2

1' Wing without macrotrichia on wing surface or a few present on membrane on apical portion of wing and distal portion of M1, M2, CuA1 (Figs 2A,12A,I,17A); macrotrichia never present on basal half of wing (Fig. 12I) 7

2. Gonostylus divided (Figs 7D,F,9D-E) 3

2' Gonostylus with single, undivided apex (Figs 3E-H,10E-F,11C,L) . 4

3. Wing length > 0.9 mm; gonostylus with three distinct apical teeth, appearing cuspate in some (Figs 7D-F) . . . . *Culicoides filipalpis* Remm (p. 53)

3' Wing length = 0.6 mm; gonostylus with subapical lobe projecting dorsomedially (gonostylus appearing apically bifurcate from some angles) (Figs 9D-E) . . . . *Culicoides tyrrelli* (Boesel) (p. 55)

4. Antenna with terminal four flagellomeres elongate (Fig. 3B), antennal flagellomere 10/11 > 0.84; palpal segments 3/4 >1.54; gonostylus slender (Figs 3E-H) . . . . . *Atriculicoides globosus* (Boesel) (p. 39)

4' Antenna with terminal three flagellomeres elongate (Figs 10B,11B,11I), antennal flagellomere 10/11 < 0.45; palpal segments 3/4 <1.62; gonostylus thick (Figs 10E-F,11C,L) 5

5. Wing length < 0.68 mm; wing with macrotrichia mostly restricted to apical half of wing (Fig.10A); gonostylus pointed apically (Figs 10E-

F) . . . . *Culicoides canadensis* (Boesel) (p. 58)

5' Wing length > 0.68 mm; wing with macrotrichia scattered over most of wing (Figs 11A,H); gonostylus rounded apically (Figs 11C-L) . 6

6. Gonostylus short, thick, of even thickness along length (Fig. 11C); apicolateral process very small, pointed (Fig. 11C) . . . . *Culicoides bullus* n.sp. (p. 61)

6' Gonostylus more elongate, enlarged basally (Fig. 11L); apicolateral process elongate (Fig. 11L) . . . . . *Culicoides agamus* n.sp. (p. 62)

7. Costal ratio ≤ 0.36 (Figs 1B,13B); antennal flagellomeres of nearly equal length, except for last which is longer than preceding flagellomeres; palpus with one segment beyond palpal segment 3 (sensilla bearing segment) 8

7' Costal ratio ≥ 0.43; at least last three antennal flagellomeres elongate; palpus with two segments beyond palpal segment 3 (sensilla bearing segment) 9

8. Terminal antennal flagellomere elongate, enlarged basally (Fig. 1C); midtibia with apical spur; tergite 9 with posteromedial projections (Figs 1D-E) . . . . *Leptoconops primaevus* n.sp. (p. 30)

8' Terminal antennal flagellomere oblong (Fig. 13E); midtibia without apical spur; tergite 9 without posteromedial projection (Fig. 13D) . . . . *Minyohelea pumilis* n.sp. (p. 69)

9. Wing length ≥1.16 mm; costal ratio ≥ 0.80 . . . . *Protoculicoides depressus* Boesel (p. 34)

9'. Wing length ≤0.93 mm; costal ratio ≤ 0.62 10

10. Wing without macrotrichia on membrane (Figs 6H,14K); or, if present, restricted to very apical margin of wing (Fig. 16A) 11

10' Wing with at least some macrotrichia on membrane on apical portion of wing and along at least distal portion of M1 and M2 (Figs 12A,I, 17A) 14

11. Palpal segments 3/4 ≥ 1.50; hind tibia with dorsal row of elongate setae thick (Fig. 6K) . . . . *Heleageron arenatus* n.sp. (p. 48)

11' Palpal segments 3/4 ≤ 1.33; hind tibia with dorsal row of elongate setae slender 12

12. Wing length = 0.45 mm; gonostylus thick (Fig. 6B); hindleg first tarsomere without thick basal spine and palisade setae (Fig. 6G) . . . . *Adelohelea glabra* n.sp. (p. 46)

12' Wing length ≥ 0.55 mm; gonostylus with pointed apex (Figs 12C,F,M-N,14L,O,16B,F,17C-E); hindleg first tarsomere with thick basal spine and row of palisade setae 13

13. Gonostylus straight or slightly curved (Figs 16B,F) . . . . *Palaeobrachypogon vetus* n.sp. (p. 78)

13' Gonostylus strongly curved at midlength (Figs 14L,O) . . . . *Palaeobrachypogon remmi* n.sp. (p. 76)
14. Hind first tarsomere with thick basal spine (Figs 18C-D) . . . . *Peronehelea chrimikalydia* n.sp. (p. 82)
14' Hind first tarsomere without thick basal spine 15
15. Gonostylus with inner lobe at midlength (Fig. 12C) . . . . *Culicoides obuncus* n.sp. (p. 64)
15' Gonostylus without inner lobe (Figs 12M-N) . . . . *Culicoides annosus* n.sp. (p. 65)

Females

The females of *Culicoides agamus, C. obuncus*, and *C. annosus* are unknown.

1. Hindleg claws elongate with hind claw length/hindleg fifth tarsomere length > 0.8 (Figs 15I,M,16L,18G-I) 2
1' Hindleg claws moderately developed or short with hind claw length/ hindleg fifth tarsomere length < 0.6 (Figs 10N,13L, 14H) 4
2. Hindleg claw a single talon with a slender inner tooth (Figs 18G-I). . . . . *Peronehelea chrimikalydia* n.sp. (p. 82)
2' Hindleg claw paired, equal or subequal (Figs 15I,M,16L) 3
3. Wing without macrotrichia on membrane (Fig. 16H); hindleg claws elongate (Fig. 16L), at least 1.5x the length of the fore and midleg claws (Figs 16J-K) . . . . . . *Palaeobrachypogon vetus* n.sp. (p. 78)
3' Wing with macrotrichia along apical margin of wing (Fig. 15A); hindleg claws of moderate length (Figs 15I,M), similar in size to fore and midleg claws (Figs 15J-L) . . . . *Palaeobrachypogon remmi* n.sp. (p. 76)
4. Wing with macrotrichia present on most of membrane, including on basal half of wing and in cells cua_1 and cua_2 (Figs 4C, 5C, 8A, 9F,10H,11D) 5
4' Wing without macrotrichia on membrane of wing surface (Figs 1F,2E,6D,M,13F) or with macrotrichia only along apical margin and none in cells cua_1 and cua_2 (Fig. 14A) 10
5. Costal ratio > 0.73 . . . . *Atriculicoides globosus* (Boesel) (p. 39)
5' Costal ratio < 0.72 6
6. Mouthparts elongate (length/length of foreleg fifth tarsomere > 3.5) 7
6' Mouthparts of moderate length (length/length of foreleg fifth tarsomere < 3.6) (Figs 8G,9L) 8
7. Antennal ratio >1.08; palpal segment 3 swollen (Fig. 9L) . . . . *Culicoides tyrrelli* (Boesel) (p. 55)
7' Antennal ratio <1.19; palpal segment 3 slender and elongate (Fig. 8G,K) . . . . *Culicoides filipalpis* Remm (p. 53)

8. Third palpal segment slender (Fig. 10J); flagellomeres 1-8 elongate (Fig. 10K) . . . . *Culicoides canadensis* (Boesel) (p. 58)

8' Third palpal segment swollen (Figs 5G-H,11E); flagellomeres 1-8 squat to globular (Fig. 5E,11G) 9

9. Wing length < 0.7 mm; costal ratio < 0.64 . . . . *Culicoides bullus* n.sp. (p. 61)

9' Wing length > 0.8 mm; costal ratio > 0.65 . . *Atriculicoides* sp. (p. 43)

10. Costal ratio < 0.45 (Figs 1F,13F); antennal flagellomeres of nearly equal length, except for last which may be longer than preceding flagellomeres (Figs 1G,13H); palpus with one segment beyond palpal segment 3 (sensilla bearing segment) (Figs 1K, 13G) . . . 11

10' Costal ratio > 0.54; at least last three antennal flagellomeres longer than preceding flagellomeres (Figs 2G,6E,N,14I); palpus with two segments beyond palpal segment 3 (sensilla bearing segment) . 12

11. Eyes separated dorsomedially by large gap (Fig. 1K); last flagellomere longer than penultimate flagellomere (Fig. 1G); radial veins of wing compacted anteriorly (Fig. 1F); cerci very elongate (Figs 1L-M) . . . . *Leptoconops primaevus* n.sp. (p. 30)

11' Eyes separated dorsomedially by narrow gap (as in Fig. 2H); last two flagellomeres equal in size (Fig. 13H); radial veins of wing short but not compacted anteriorly (Fig. 13F); cerci short (Fig. 13I) . . . . *Minyohelea pumilis* n.sp. (p. 69)

12. Wing length = 1.6 mm; costal ratio > 0.80 . . . . *Protoculicoides depressus* Boesel (p. 34)

12' Wing length < 0.9 mm; costal ratio < 0.74 13

13. Wing with single radial cell (Fig. 6M) . . . . *Heleageron arenatus* n.sp. (p. 48)

13' Wing with two radial cells (Figs 6D,14A) 14

14. Wing with macrotrichia on apical margin (Fig.14A); hind first tarsomere with strong basal spine and palisade setae (Fig. 14D); claws of moderate size (Figs 14C-E,G,H) . . . . *Palaeobrachypogon aquilonius* (Boesel) (p. 74)

14' Wing without macrotrichia on membrane; hind first tarsomere without strong basal spine or palisade setae (Fig. 6G); claws small (Figs 6F-G) . . . . *Adelohelea glabra* n.sp. (p. 46)

Unique or distinctive character states useful for identifying Canadian amber Ceratopogonidae

The following features are distinctive character states useful for identifying some species or genera of Ceratopogonidae in Canadian amber.

- Eyes broadly contiguous dorsomedially (Figs 4H, 5D): *Atriculicoides*.
- Antenna with only terminal flagellomere elongate (Figs 1C,G;13E):

Leptoconops primaevus, male *Minyohelea pumilis* (the latter is much smaller than *L. primaevus*).
- Antenna with all flagellomeres of nearly equal length (Fig. 13H): female *Minyohelea pumilis.*
- Antenna with 11 flagellomeres (Fig. 6C): male *Adelohelea glabra.*
- Antenna with 12 flagellomeres (Figs 1C,G): *Leptoconops primaevus.*
- Palpus with only one segment beyond third (that bearing capitate sensilla) (Figs 1K;13C,G): *Leptoconops primaevus, Minyohelea pumilis* (the latter is much smaller than *L. primaevus*).
- Palpus with very elongate segments (Fig. 8K): female *Culicoides filipalpis.*
- Wing with short radial sector (Figs 1B,F; 13B,F): *Leptoconops primaevus, Minyohelea pumilis* (the latter is much smaller than *L. primaevus*).
- Female wing with long radial sector (Fig. 2A,E) (Costal ratio =0.89): *Protoculicoides depressus.*
- First tarsomere of foreleg with series of stout bristles (Fig. 3C): *Atriculicoides globosus.*
- Female claws of all legs well developed and paired (Figs 14G, H; 15I-M; 16J-L): *Palaeobrachypogon.*
- Claws of hindleg paired and substantially longer than those of fore and midleg (Figs 16J-L): *Palaeobrachypogon vetus.*
- Hindleg with claw a single elongate talon (and short basal tooth) (Figs 18G-I): *Peronehelea chrimikalydia.*
- Gonostylus with tricuspate apex (Figs 7D,F): *Culicoides filipalpis.*
- Gonostylus with forked apex (Figs 9D-E): *Culicoides tyrrelli.*
- Female cerci very elongate (Figs 1L-M): *Leptoconops primaevus.*

DESCRIPTION OF TAXA IN CANADIAN AND NEW JERSEY AMBER

Leptoconops Skuse

Leptoconops Skuse, 1889:288. Type-species, *Leptoconops stygius* Skuse, by monotypy.

Diagnosis and Description of Adults. As given by Clastrier and Wirth (1978), Smee (1966), Szadziewski (1988:232), and Wirth and Atchley (1973). In addition, three anterior scutal apodemes present; anapleural suture well developed, extending to anterior margin of anepisternal cleft.

Distribution and bionomics. Extant members of the genus *Leptoconops* are found worldwide but appear to be generally restricted to drier habitats in tropical, subtropical, and warm temperate climatic regions. Species of *Leptoconops* have also been found as far north as Moscow, Russia (Szadziewski, 1988) and the southern Yukon Territory, Canada (Downes, 1983) in the Old and New World, respectively.

Immatures are known from moist sand bordering marine, saline or freshwater habitats or, more rarely from clay, or halomorphic soils. Extant adult females are diurnal feeders and are known to take blood meals from a variety of vertebrate hosts, including lizards (Auezova et al., 1990), birds (Lee, et al., 1962; Mayer, 1955), and mammals (Lee et al., 1962; Mayer, 1955; Wirth and Atchley, 1973). Two species are reported to be at least partially autogenous (Linley, 1968; Rees et al., 1971).

Taxonomic discussion. The 122 extant species of *Leptoconops* are placed in six subgenera (with number of extant species in parentheses): *Styloconops* Kieffer (8), *Brachyconops* Wirth and Atchley (1), *Megaconops* Wirth and Atchley (1), *Leptoconops* (*L.*) (52), *Holoconops* Kieffer (53) and *Proleptoconops* Clastrier (7).

Two fossil species have been previously described and named within the genus (Table 18), one from Baltic amber (Eocene-Lower Oligocene) (Szadziewski, 1988) and the other from the Upper Cretaceous of the Taimyr Peninsula in Russia (Kalugina, 1991). Szadziewski (1990a) described but did not name a female from Sakhalin amber, of unspecified Tertiary age. Schlüter (1978:100) provided a brief description of a female *Leptoconops* from Upper Cretaceous amber in France (also discussed by Szadziewski and Schlüter, 1992). Finally, Grimaldi (1993a) included a photograph of what is clearly a female *Leptoconops* from Upper Cretaceous New Jersey amber.

Grogan and Szadziewski (1988) suggested that the Lower Cretaceous fossil *Simulidium priscum* Westwood (1854:394), known as a single wing, is likely a member of *Leptoconops.* Long considered lost, the fossil has been located in the BMNH and is actually a member of the Brachycera (Ross, pers. comm.).

Of the two previously named fossil species of *Leptoconops, L. boreus* Kalugina is known only as a female, while both the male and female of *L. succineus* Szadziewski are described. The following keys will distinguish between these and the new species described below.

Key to male fossil *Leptoconops*

1. Apicolateral process moderately elongate, broad; inner margin of gonocoxite without elongate setae; outer margin of gonostylus with short seta. . . .*succineus* (Baltic amber, Eocene-Lower Oligocene)
1' Apicolateral process very elongate, slender; inner margin of gonocoxite with elongate setae; outer margin of gonostylus with elongate seta (Figs 1D-E) . . . .*primaevus* (Canadian amber, Upper Cretaceous)

Key to female fossil *Leptoconops*

1. Antennal flagellomeres 7-11 oval. . . . . *succineus* Szadziewski (Baltic amber, Eocene-Lower Oligocene)
1' Antennal flagellomeres 7-11 squat or spherical 2
2. Antennal flagellomeres 7-11 squat . . . .*boreus* Kalugina (Taimyr amber, Upper Cretaceous)
2' Antennal flagellomeres 7-11 spherical (Fig. 1G) . . . .*primaevus* n.sp. (Canadian amber, Upper Cretaceous)

Leptoconops (Leptoconops) primaevus Borkent n. sp.

Diagnosis. Male: only Cretaceous species of Ceratopogonidae with only the terminal flagellomere elongate. Female: only Cretaceous species of Ceratopogonidae with a midleg tibial spur and with antennal flagellomeres 7-11 spherical.

Description. Male adult. Descriptive statistics: see Tables 3-8.

Head: Eyes bare, broadly separated dorsomedially. Vertex without single dorsomedial seta. Antenna with well developed plume, basal foramen of pedicel not visible, 13 separate flagellomeres, flagellomere 13 more elongate than preceding flagellomeres (Fig. 1C). Mouthparts moderately elongate. Palpus with four segments, third segment somewhat ovoid, with capitate sensilla arranged on surface.

Thorax: Anterior scutal apodemes not visible. Scutum pruinose, with a few scattered elongate setae, humeral pit not visible. Anapleural suture not visible. Katepisternum with several setae.

Wing (Fig. 1B): Without macrotrichia, fine microtrichia on all membrane. Alula without macrotrichia. Cells r_1 and r_{2+3} compacted anteriorly, not distinguishable. Base of M poorly defined, apparently bifurcating distal to apex of r_{2+3}.

Legs: Femora, tibiae slender. Legs lacking armature except apex of first and second tarsomeres of each leg with pair of thick setae. Fore and midleg trochanter each without pair of thick setae. Midleg tibia with apical spur. Hindleg first tarsomere without thick basal spine, without palisade setae. Claws simple. With short empodium.

Genitalia (Figs 1D-E): Apicolateral processes broadly separated, each elongate, slender, with apical seta. Gonocoxite moderately elongate. Gonostylus thick basally, tapering to toothed apical spine, bearing thick seta on outer face at midlength. Paramere semiglobular apically. Aedeagus not visible.

Female adult. Descriptive statistics: see Tables 9-14.

Head: Eyes bare, broadly separated dorsomedially. Vertex without single dorsomedial seta (Fig. 1K). Antennae with 12 separate flagellomeres, terminal flagellomere more elongate than preceding flagellomeres (Fig. 1G), first flagellomere sensilla not visible. Mouthparts moderately elongate (Fig. 1K). Clypeus separated laterally by membrane from head capsule. Labrum not visible. Mandible with fine teeth. Lacinia with at least six, large, retrorse teeth. Palpus with four segments, third segment somewhat ovoid, with capitate sensilla arranged on surface (Fig. 1K). Hypopharynx not visible.

Thorax: Anterior scutal apodemes not visible. Scutum pruinose, with a few scattered elongate setae, humeral pit not visible. Anapleural suture not visible. Katepisternum with several setae.

Wing (Fig. 1F): Without macrotrichia, fine microtrichia present on all membrane. Alula without macrotrichia. Cells r_1 and r_{2+3} compacted anteriorly, not distinguishable. Base of M poorly defined, apparently bifurcating distal to apex of r_{2+3}.

Legs: Femora, tibiae slender. Legs lacking armature, except for pair of thick setae on apices of at least first and second tarsomeres of all legs, in some specimens additional thick setae arranged ventrally on first tarsomere of each leg. Foreleg, midleg, hindleg claws equal, simple, with very slender inner tooth arising from near base of each talon (Figs 1H-J). Hindleg claw shorter than fifth tarsomere. Claws of all legs of equal length. Hind first tarsomere without thick basal spine, without palisade setae. With short empodium.

Genitalia (Figs 1L-M): Spermathecae not visible. Posterior margin of sternite 8 apparently without thick setae. Sternite 9 not visible. Segment 10 with pair of setae. Cerci elongate, laterally compressed.

Distribution and bionomics. L. primaevus is known from amber collected 5 km south of Grassy Lake, Alberta (n= 1♂, 6 ♀) and from the shore of Cedar Lake, Manitoba (1 ♂, 2 ♀) (Fig. 20B).

The immatures of most extant species of *Leptoconops* are known only from sand or sandy soil which include some salts (Linley, 1976; Matayoshi et al., 1985; Wirth and Atchley, 1973). Rioux et al. (1968), however, found larvae of *L. irritans* (Noè) in France in calcareous, halomorphic soil which was susceptible to periodic drying, Magnon et al. (1990) recorded *L. linleyi* Wirth and Atchley, a species normally associated with sandy soils, from salt marsh soils (but in relatively low numbers) and Smith and Lowe (1948) observed that *L. carteri* Hoffman (as *L. torrens* Townsend) larvae in wet clay soils in California which were subjected to seasonal drying and had a salt concentration of 400 parts per million.

Because *L. primaevus* belongs to a relatively derived clade within the genus (see "Phylogenetic Relationships of Cretaceous Ceratopogonidae"), these habitat restrictions suggest that the immatures of *L. primaevus* were also likely present in sand or sandy soils, as recorded for most extant species of *Leptoconops*. Further evidence supports this conclusion, as follows.

Most females of species in the genus (those in the subgenera *Leptoconops, Holoconops, Megaconops* and *Proleptoconops*) are unique in the Ceratopogonidae (and in the Culicomorpha) in having very elongate, laterally compressed cerci. It seems likely that these are used primarily to assist in laying eggs in sand or sandy soil. The only observations of oviposition in the genus (Rees et al., 1971) indicate that females of the *L. kerteszi* complex insert their eggs into sand. All other Culicomorpha either lay their eggs on the surface of substrates or water or drop their eggs into water. If the elongate, compressed cerci are indeed functionally related to oviposition in sand or sandy soils, the presence of such cerci in *L. primaevus* would further support their association with sandy habitats. Finally, as described below under 'Palaeoecology', it is clear that sandy habitats were very common in Alberta during the Upper Cretaceous and this suggests that the sandy banks of the meandering rivers known to be present at that time (some possibly experiencing marine tidal incursions; see Fig. 20A) was the likely breeding habitat of *L. primaevus*.

It is striking that *L. primaevus* is relatively common, representing 6.29% of all Ceratopogonidae in Canadian Cretaceous amber (Table 1). In contrast, Szadziewski (1988) reported that *Leptoconops* made up only 0.36% of the 1,103 Ceratopogonidae in Baltic amber. In addition, the presence of a male and female in one piece of amber and two females in another, also indicates that *L. primaevus* was likely common, at least locally. Finally, as noted above, sand deposits are abundant in the Upper Cretaceous of Alberta and these likely represent the breeding sites of *L. primaevus*. All these observations indicate that *L. primaevus* was likely an abundant component of the Ceratopogonidae fauna of southern Alberta during the Upper Cretaceous.

Female *L. primaevus* likely fed on the blood of unidentified vertebrate hosts (see below under "Evolution of blood-sucking and host selection in Ceratopogonidae").

One female of *L. primaevus* was associated with a female of *Atriculicoides globosus* in a single piece of amber.

Taxonomic discussion. Males and females were associated through the presence of a male and a female in a single piece of amber from Cedar Lake, Manitoba. In addition, the specimens were the only members of this distinctive genus in the amber.

The paratype male differed somewhat from the holotype (Figs 1D-E). The single seta on the outer margin of the gonostylus appeared thick and some of the claws (but not all) appeared apically bifid. This may indicate the presence of two species but I was reluctant to recognize these. It is possible that these apparent differences may be an artifact of preservation. Other setae on the body and legs of the paratype male also appeared thick but such setae appeared in an asymmetrical pattern. These may have been coated with a foreign substance to give them this appearance. Similarly, the apparently bifid claws were asymmetrical, simple on one fore and midleg and bifid on the others. Such bifid claws would, furthermore, be unique within the genus (see "Phylogenetic Relationships of Extant Ceratopogonidae" synapomorphy 11). Further material is needed to test these potential differences.

Types. Holotype, male adult in amber on microscope slide, labelled "Holotype Leptoconops primaevus Borkent, nr. Medicine Hat [= 5 km s. Grassy Lake], Alta., 1973, P. Boston, CAS 792, CNC No. 21659" (CNCI); allotype, female adult in amber on microscope slide, labelled "nr. Medicine Hat, Alta., May, 1973, McAlpine & Teskey, CAS 733" (CNCI); paratypes: 1 male, 1 female (in one piece of amber), Cedar Lake, Manitoba, F.M.C. Collection, MCZ 9001 (MCZC); 1 female, Cedar Lake, Manitoba, F.M.C. Collection, MCZ 9000 (MCZC); 2 females, near Medicine Hat [= 5 km s. Grassy Lake], Alberta, July 8-11, 1971, McAlpine, CAS 335, CAS 462 (CNCI); 1 female (in one piece of amber with a female of *Atriculicoides globosus*), near Medicine Hat [= 5 km s. Grassy Lake], Alberta, May, 1973, McAlpine & Teskey, CAS 614a (CNCI); 2 females (one incomplete, in one piece of amber), near Medicine Hat [= 5 km s. Grassy Lake], Alberta, June, 1974, McAlpine, Cooper & Daze, CAS 929 (CNCI).

Derivation of specific epithet. The name *primaevus* (early) refers to the age of this Upper Cretaceous species.

Protoculicoides Boesel

Protoculicoides Boesel, 1937:50. Type species *Protoculicoides depressus* Boesel, by original designation.

Diagnosis. Male: only Cretaceous Ceratopogonidae with a CR = 0.80 (all others with CR ≤ 0.62). Female: only Cretaceous Ceratopogonidae with CR = 0.89 and a wing without macrotrichia on the membrane (all other taxa with or without macrotrichia, CR ≤ 0.74).

Description: Moderate size adults. Eye pubescence not visible, eyes narrowly separated dorsomedially. Vertex with single dorsomedial seta present, with long setae all along vertex (Fig 2H). Male antennae with 13 separate flagellomeres, flagellomeres of nearly equal length except for more elongate flagellomeres 10-13 (Fig. 2C), sensilla coeloconica not present, plume well developed, basal foramen of pedicel not visible. Female antenna with 13 separate flagellomeres, flagellomeres of nearly equal length except for more elongate flagellomeres 10-13 (Fig. 2G), sensilla coeloconica not visible. Details of female mouthparts not visible (Fig. 2I). Palpus with 5 segments, pit possibly present in male (Fig. 2B), not present in female (Fig. 2I). Anterior scutal apodemes not visible. Female scutum with numerous long setae scattered over surface, humeral pits well developed. Male scutum with numerous long setae, pattern not visible, humeral pits not visible. Anapleural suture not visible. Katepisternal setae not visible. Legs slender, unarmed except for pair of thick bristles at apex of tarsomeres 1-4 of each leg. Midleg trochanter with pair of thick setae in male, not visible in female. Midleg tibia without apical spur. Apex of hindleg tibia with 4-5 elongate setae, spur present in male, not visible in female. Hindleg first tarsomere straight, with scattered setae along entire length, without basal thick spine. Female claws equal, short, without inner teeth. Male claws apically simple. Empodium present, moderately simple. Male wing with fine microtrichia, macrotrichia restricted to apex of wing, apical portions of M_1, M_2, alula with macrotrichia, cells r_1 and r_{2+3} well defined, elongate, M bifurcating distal to r-m (Fig. 2A). Female wing with fine microtrichia, without macrotrichia on membrane, alula with macrotrichia, cells r_1 and r_{2+3} well defined, elongate, M bifurcating distal to r-m (Fig. 2E). Male genitalia (Fig. 2D) moderate in size, with broad, tapering apicolateral process, apical seta not visible. Female genitalia with spermathecae not visible, sternite 9 not visible but without vaginal apodeme, cercus of moderate size.

Taxonomic discussion: Protoculicoides was placed in the Culicoidini by Szadziewski (1988) but removed from this tribe by Szadziewski and Schlüter (1992). Its cladistic position is discussed below under "Phylogenetic Relationships of Cretaceous Ceratopogonidae".

Protoculicoides depressus Boesel

Protoculicoides depressus Boesel, 1937:51. Cedar Lake, Manitoba. Upper Cretaceous.

Diagnosis. Male and female: as for genus as this is the only included species.

Description. Male adult. As for genus as this is the only included species. Descriptive statistics: see Tables 3-8.

Genitalia (Fig. 2D): Tergite 9 with well developed apicolateral process, setae on apicolateral process not visible. Paramere elongate, simple, slightly sinuous. Aedeagus with well developed basal apodemes, a single extension posteriorly. Gonocoxite moderately elongate. Gonostylus elongate, of uniform diameter, slightly bent apically.

Female adult. As for genus as this is the only included species. Descriptive statistics: see Tables 9-14.

Distribution and bionomics. Protoculicoides depressus is known from amber collected from the shore of Cedar Lake, Manitoba (holotype female) and from 5 km south of Grassy Lake, Alberta (n= 2 ♂) (Fig. 20B).

The details of the mouthparts of the female of *P. depressus* were not visible and therefore its feeding habits cannot be deduced using these (see below under "Evolution of blood-sucking and host selection in Ceratopogonidae"). Phylogenetic relationships provide no further clues of the likely mode of feeding.

One of the pieces of amber with a male *P. depressus* (CAS 1015) also included a female *C. filipalpis*.

Taxonomic discussion. The males and females were associated through the shared presence of several features: elongate costa; female wing without macrotrichia and the male with only a few at the wing apex; presence of similar armature of the legs (pair of stout bristles at the apex of tarsomeres 1-4 of each leg); and the presence of numerous long setae on the scutum.

The female holotype of *P. depressus* was removed from the original slide mount in Canada Balsam and the amber further cut and polished. The amber piece was mounted in several different positions, allowing for a more detailed examination of the specimen. Although it was not labelled as such, it is clear the holotype was collected from Cedar Lake, Manitoba, where all early Canadian amber originated (including the other species which Boesel (1937) described).

Boesel (1937:51) noted that the thorax of the female type was depressed and suggested that this might not be natural. It is clear that the specimen is indeed flattened dorsoventrally; the head capsule, antennal pedicel, anterior portion of the thorax (including the cervical area) and the abdomen are all strikingly flattened, in a slightly asymmetrical manner. These observations indicate that the flattened appearance is the result of compression. In addition, if the males and female are correctly associated, this would be further evidence that the compression is an artifact of preservation as the males have normally shaped, somewhat rounded thoraces.

Boesel (1937:54) also figured the wing of *P. depressus* but did not draw the anal portion of the wing because this could not be seen. By reworking the amber piece, the entire wing could be clearly observed and reported here.

Szadziewski (1988:28) reported that *P. depressus* has pubescent eyes but this observation could not be confirmed. A few areas of ommatidia could be clearly seen in outline and these lacked any evidence of spicules.

Walker (1934:pl.1, fig. 6) provided a photograph of *P. depressus* (identified as "Biting midge").

Type. Holotype, female adult in amber on microscope slide, labelled "Holotype Protoculicoides depressus "Boesel"", "Dept. of Entomology, ROYAL ONTARIO MUSEUM, Toronto, Canada", "6" [on back of slide] (ROME).

Material examined. 2 males: CAS 1015 (CNCI), CAS 1231 (CNCI); 1 female: holotype ROM 6 (ROME).

Derivation of specific epithet. The name *depressus* given by Boesel (1937) certainly refers to the flattened dorsal surface of the scutum of the holotype which Boesel (1937) noted might be an artifact.

Atriculicoides Remm

Atriculicoides Remm 1976:108. Type species *Atriculicoides macrophthalmus* Remm, by original designation.

Diagnosis. Male and female: only Ceratopogonidae with eyes broadly contiguous, a terminal flagellomere tapering gradually to its apex (no basally constricted nipple present) and a hind tibial spur.

Description: Moderate size adults, most of body with numerous long setae. Eye pubescence present or absent, eyes broadly contiguous dorsomedially. Vertex with single dorsomedial seta present, with long setae along vertex (Figs 4H,5D). Male antennae with 13 separate flagellomeres, flagellomeres of nearly equal length except for more elongate flagellomeres 10-13 (Fig. 3B), sensilla coeloconica not present, plume well developed, pedicel with basal foramen of intermediate size. Female antenna with 13 separate flagellomeres, flagellomeres of nearly equal length except for more elongate flagellomeres 9-13 (Figs 4D,5E), sensilla coeloconica not present. Proboscis short (Fig. 4H), labrum with fine teeth, mandibles with fine teeth, lacinia with retrorse teeth (Fig. 4J). Palpus with 4-5 segments, if four present, only one present distal to segment 3 (bearing the capitate sensilla), segment 3 with capitate sensilla scattered on surface, in broad, shallow depression, or in deep pit (Figs 4I,5G-H). Hypopharynx not visible. Anterior scutal apodemes not visible. Scutum with numerous setae of varying lengths densely scattered over surface, humeral pits present, of moderate size or absent. Anapleural suture short, extending to posterior margin of anepisternal cleft (Fig. 3D). Katepisternal setae absent. Legs slender, unarmed or with a few stout bristles present on first tarsomere of both foreleg, hindleg. Midleg trochanter with pair of thick setae. Midleg tibia without apical spur. Apex of hindleg tibia with 4-6 elongate setae, spur present, simple (Figs 4G,5O). Hindleg first tarsomere straight, with scattered setae along entire length, without basal thick spine (Figs 4G,5O). Female claws equal, short, with large or very slender, sliver-like inner teeth (Figs 4K-M,5I-L). Male claws apically bifid. Empodium present or absent. Wing with fine microtrichia, macrotrichia present over entire wing in female (Figs 4C,5C), present over entire wing (Fig. 3A) or restricted to apical portion of wing (in cells r_5, m_1, m_2 at wing apex) in male, alula with macrotrichia, cells r_1 and r_{2+3} well defined, elongate, M bifurcating distal to r-m, base of M_2 present or absent. Male genitalia moderate in size, with apicolateral process (Figs 3E-H,4A-B). Female genitalia with at least 2 semispherical spermathecae (Fig. 5A), sternite 9 medially continuous, without vaginal apodeme, cercus of moderate size (Figs 5A-B).

Distribution and bionomics. Species of *Atriculicoides* are known from four Upper Cretaceous sites: France, the Taimyr Peninsula in Russia and the two sites in Canada. Some bionomic information, deduced from material of *A. globosus*, is discussed below under that species.

Species of *Atriculicoides* were relatively common in Canadian amber, making up 29.56% of all Ceratopogonidae (Table 1).

Szadziewski and Schlüter (1992) suggested that the presence of unmodified legs, short and equal claws, rudimentary empodia and numerous sensilla on the third palpal segment indicate that female *Atriculicoides* may have fed on vertebrate blood. Observations of the mouthparts of the two species recognized here confirm this hypothesis (see below under "Evolution of blood-sucking and host selection in Ceratopogonidae"). The specific type of vertebrate host, however, is uncertain.

Taxonomic Discussion: There are four previously recognized species of *Atriculicoides*, all from the Upper Cretaceous; two of these are from Taimyr amber, *A. macrophthalmus* Remm and *A. squamiciliatus* Remm (Remm, 1976) and two from French amber, *A. cenomanensis* Szadziewski and Schlüter and *A. incompletus* Szadziewski and Schlüter (Szadziewski and Schlüter, 1992). The transfer of *A. globosus* here makes a total of five species known in the genus (Table 18).

The only reported difference between *A. macrophthalmus* and *A. squamiciliatus* are small differences in the shape of the scales on the fringe of the wing (Remm, 1976). These differences may be the result of an artifact of preservation as the scales of numerous specimens examined in this study had some striking modifications which were clearly due to a layer of gas or contamination. Furthermore, there is some confusion about the actual character states present in each of these two species. In Remm (1976), his figure 1a shows the wing of *A. macrophthalmus* with narrow wing scales and his figure 1f shows the wing of *A. squamiciliatus* with broader scales. However in the text, Remm (1976) states that *A. macrophthalmus* has a "thick fringe [translated from Russian]" while *A. squamiciliatus* has a "fringe consisting of narrow scales [translated from Russian]". The key below has assumed that the figures accurately depict the character states. Clearly, it would be invaluable to restudy the original material to examine these and any other possible differences.

Remm (1976) reported the presence of pubescence on the eyes of his two Taimyr amber species *A. macrophthalmus* and *A. squamiciliatus* and Szadziewski and Schlüter (1992) reported the eyes of *A. cenomanensis*, from French amber, were "probably pubescent". The specimens of *A. globosus* examined here had eyes devoid of pubescence. However, it should be noted that this character was very difficult to observe accurately within the amber and a number of specimens which initially appeared to have some pubescence, upon further cutting and polishing of the amber, were found to lack hairs. Contamination on the surface of the specimen and distortions of light through amber may give the false impression of the presence of pubescence where actually none exist.

Szadziewski and Schlüter (1992) reported that the male of *A. cenomanensis* had palisade setae on the first tarsomere of the hindleg. This character state is incongruous with the observations of the genus made here. Unfortunately, original material of *A. cenomanensis* could not be studied firsthand to examine the possibility that the angle of observation or poor visibility, resulted in a misinterpretation of this feature (Szadziewski and Schlüter (1992) note that the specimen was poorly preserved).

The report of *Forcipomyia* Meigen in Canadian amber by Downes and Wirth (1981) actually refers to the species described here as members of *Atriculicoides* (as indicated by some notes on the specimens studied from the CNCI).

Key to Fossil *Atriculicoides*

The male of *A. incompletus* and the female of *A. cenomanensis* are unknown.

1. Palpus with four segments, only one segment distal to segment 3 (bearing the capitate sensilla) . . . .
   . . . *incompletus* Szadziewski and Schlüter (French amber, Upper Cretaceous)

1' Palpus with five segments (Figs 4H,I,5G,H) 2

2 First tarsomere of foreleg with row of stout bristles (Fig. 3C). 3

2' First tarsomere of foreleg with scattered, slender setae 4
3. Palpal segment 3 without pit, capitate sensilla scattered on surface of segment or in shallow depression (Fig. 4I) . . . . *globosus* (Boesel) (Canadian amber, Upper Cretaceous)
3' Palpal segment 3 with well defined pit (Fig. 5G,H) . . . . *Atriculicoides* sp.
4. Male 5
4' Female 7
5. Palpal segment 3 with sensory pit; wing with macrotrichia restricted to apical portion of wing, present in cells r_5, m_1, m_2 . . . . *cenomanensis* Szadziewski and Schlüter (French amber, Upper Cretaceous)
5' Palpal segment 3 with sensory sensilla scattered on surface or in shallow depression; wing with macrotrichia present over entire wing 6
6. Scales on posterior wing fringe slender. . . . . *macrophthalmus* Remm (Taimyr amber, Upper Cretaceous)
6' Scales on posterior wing fringe broad, thick . . . . *squamiciliatus* Remm (Taimyr amber, Upper Cretaceous)
7. Scales on posterior wing fringe slender . . . . *macrophthalmus* Remm (Taimyr amber, Upper Cretaceous)
7' Scales on posterior wing fringe broad, thick . . . . *squamiciliatus* Remm (Taimyr amber, Upper Cretaceous)

Derivation of generic name: The name *Atriculicoides* given by Remm (1976) refers to his recognition that this fossil genus combines characteristics of extant members of *Atrichopogon* and *Culicoides*.

Atriculicoides globosus (Boesel) NEW COMBINATION

Lasiohelea globosa Boesel, 1937:47. Cedar Lake, Manitoba. Upper Cretaceous.
Lasiohelea cretea Boesel, 1937:46. Cedar Lake, Manitoba. Upper Cretaceous. NEW SYNONYM, NEW COMBINATION.

Diagnosis. Male and female: only species of *Atriculicoides* with palpal segment 3 without a pit and with thick spines on first tarsomere of foreleg.

Description. Male adult. Descriptive statistics: see Tables 3-8.

Head: Eyes bare, broadly contiguous dorsomedially. Vertex with single dorsomedial seta present. Antenna with well developed plume, 13 separate flagellomeres, basal foramen of pedicel of intermediate size, flagellomeres 10-13 more elongate than preceding flagellomeres (Fig. 3B), first flagellomere without sensilla coeloconica. Mouthparts moderately elongate. Palpus with five segments, third segment elongate, somewhat swollen near midlength, with capitate sensilla arranged on surface or in shallow depression.

Thorax: Anterior scutal apodemes not visible. Scutum with numerous setae of varying lengths densely scattered over surface, humeral pit not visible. Anapleural suture extending to posterior margin of anepisternal cleft (Fig. 3D). Katepisternum without setae.

Wing (Fig. 1A): Macrotrichia present on all membrane, fine microtrichia on all membrane. Alula with macrotrichia. Cells r_1 and r_{2+3} well defined, elongate. M bifurcating distal to r-m.

Legs: Femora, tibiae slender. Legs lacking armature, except foreleg first tarsomere with row of short, thick spines; in some specimens first tarsomere of midleg with a few short, thick spines (Fig. 3C). Fore and midleg trochanter each with pair of thick setae. Midleg tibia without apical spur. Hindleg first tarsomere without thick basal spine, without palisade setae. Claws apically bifid. Without or with short empodium.

Genitalia (Figs 3E-H,4A-B): Apicolateral process cylindrical, with single apical seta (but see taxonomic discussion below). Gonocoxite moderately elongate. Gonostylus elongate, of uniform diameter, slightly curved or straight. Parameres fused, forming single median structure with long, tapering posterior extension (but see taxonomic discussion below). Aedeagus with well developed basal apodemes, a double extension posteriorly (but see taxonomic discussion below).

Female adult. Descriptive statistics: see Tables 9-14.

Head: Eyes bare, broadly contiguous dorsomedially. Vertex with single dorsomedial seta present (Fig. 4H). Antenna with 13 separate flagellomeres, flagellomeres 9-13 more elongate than preceding flagellomeres (Fig. 4D), first flagellomere without sensilla coeloconica. Mouthparts moderately elongate (Fig 4H). Clypeus separated laterally by membrane from head capsule. Labrum with fine teeth (Fig. 4J). Mandible with fine teeth (Fig. 4J). Lacinia with at least six, large, retrorse teeth (Fig. 4J). Palpus with five segments, third segment swollen near midlength, with capitate sensilla arranged on surface or in shallow depression (Fig. 4I). Hypopharynx not visible.

Thorax: Anterior scutal apodemes not visible. Scutum with numerous setae of varying lengths densely scattered over surface, humeral pit not visible. Anapleural suture not visible. Katepisternum without setae.

Wing (Fig. 4C): Macrotrichia present on all membrane, fine microtrichia on all membrane. Alula with macrotrichia. Cells r_1 and r_{2+3} well defined, elongate. M bifurcating distal to r-m.

Legs (Figs 4E-G): Femora, tibiae slender. Legs lacking armature, except foreleg first tarsomere with row of short, thick spines; in some specimens first tarsomere of midleg with a few short, thick spines. Fore and midleg trochanter each with pair of thick setae. Midleg tibia without apical spur. Hindleg first tarsomere without thick basal spine, without palisade setae. Foreleg, midleg, hindleg claws (Figs 4K-M) equal, simple, with thick inner tooth arising from near apex of each talon. Hindleg claw (Fig. 4M) shorter than fifth tarsomere. Without empodium.

Genitalia (Figs 5A-B): Two semispherical spermathecae. Sternite 9 continuous medially, without vaginal apodeme. Segment 10 with pair of setae. Cerci of moderate length.

Distribution and bionomics. Atriculicoides globosus is known from amber collected from Cedar Lake, Manitoba (n= 6 ♂, 20 ♀) and from 5 km south of Grassy Lake, Alberta (n= 7 ♂, 9 ♀) (Fig. 20B).

Female *A. globosus* likely fed on the blood of unidentified vertebrate hosts (see below under "Evolution of blood-sucking and host selection in Ceratopogonidae").

There are a few associated organisms present in the amber containing members of *A. globosus*; with males the following were present: a Chironomidae (MCZ 9003), a male Chironomidae and a female *Culicoides tyrrelli* (MCZ 9009), and a male *Palaeobrachypogon remmi* (CAS 947). The female specimens also had associated members of other taxa: a spider (MCZ 9004), an erythraeid larval mite (CAS 768), a female *Leptoconops primaevus* (CAS 614b), a *Mesozoicaphis* Heie alate aphid and a female Chironomidae (CAS 348), and a male Chironomidae (ROM 4). The association with Chironomidae also suggests the proximity of aquatic habitats and the presence of a female *Leptoconops primaevus* indicates that *A. globosus* also bred at least near sandy areas.

A few pieces of amber held more than one specimen of *A. globosus*. A male and female were associated in each of CAS 532 and ROM 4. Two females were present in each of CAS 348, MCZ 6907 and MCZ 9007. Ten different females, or parts of different females, were present in MCZ 9008. These females were generally in a state of good preservation, were densely layered on a single plane in the amber and had overlapping legs and in some instances, had their bodies appressed against another. These observations indicate that the females of *A. globosus* likely aggregated in certain areas, perhaps resting in select places on tree trunks. Remm (1976) reported the presence of 18 females of *A. macrophthalmus* in a single piece of Taimyr amber and this may indicate that this species also aggregated as females.

The presence of associated members of the same species in five pieces of amber out of a total of 30 pieces holding at least one specimen of *A. globosus* suggests that the species was a common member of the Upper Cretaceous community in Alberta.

Taxonomic discussion. The males and females were associated through the shared presence of several features: eyes broadly contiguous dorsomedially, the presence of stout bristles on the first tarsomere of the foreleg, a wing covered with macrotrichia, and a high CR. In addition, two pieces of amber contained both a male and female of *A. globosus*; these specimens were the only members of this genus in that amber.

Poinar et al. (1993) noted four unidentified specimens of Canadian amber ceratopogonids, each with an associated parasitic mite. The specimen shown in their photograph is likely *A. globosus*, as indicated by the elongate radial cells, four elongate terminal antennal flagellomeres, the slender gonostylus and the presence of elongate apicolateral processes.

It is clear that *A. globosus* is a member of *Atriculicoides*. The presence of broadly contiguous eyes dorsomedially and a terminal antennal flagellomere without a basally constricted apical nipple indicates that the species is correctly placed.

Boesel (1937) distinguished *A. globosus* and *A. cretea* on the basis of the shape and arrangement of the basal flagellomeres. He reported that the second flagellomere of *A. cretea* was attached subapically to the first flagellomere and that flagellomeres 2-9 were wider than long. These features were evident due to the combination of observing a compressed antenna at a oblique angle. The paratype of *A. cretea* has one antenna modified as reported by Boesel (1937) but the other is typical of all other Ceratopogonidae with the second flagellomere positioned at the apex of the first. Furthermore this unmodified antenna had basal flagellomeres with a spherical to slightly elongate shape, characteristic of *A. globosus* (Fig. 4D). The two species are therefore considered to be conspecific. Boesel (1937) also noted differences in some details of the wing in *A. cretea* but these were not distinctive when more material was examined.

It is highly likely that more than one species of *Atriculicoides* is represented amongst the Canadian amber examined for this study. The males had a striking variation in the shape of the genitalia (Figs 3E-H,4A-B). However, because of differences in their state of preservation and angle of observation I was unable to satisfactorily distinguish between groups of male individuals.

Furthermore, I was unable to be certain of the identification of some of the genitalic parts of some of the males. For example, specimen MCZ 9003 appeared to be lacking apicolateral processes ((Fig. 3G) but the specimen was somewhat distorted and decomposed. In specimen CAS 631, I could not identify the aedeagus at all (Fig. 3E). In specimen ROM 41, the aedeagus appeared to come to a single apex but the genitalia were tilted laterally and a distortion on one side of the aedeagus may indicate where a second apex may have broken off (Fig. 4B). The large fused parameres present in some specimens could not be seen in others, which may indicate a specific difference. However, in none of the specimens which appeared to lack the large, well developed parameres, could I be certain that they were not merely hidden or distorted.

When further material becomes available these problems will warrant further research. For now it seems best to consider all the material as conspecific.

The pieces of amber containing the holotypes of both *A. globosus* and

A. cretea and the paratype of *A. cretea* were further cut and polished, allowing for more detailed examination of the types.

The type of *A. globosus* was chosen as the name bearer of this species because it was in a better position in the amber for study (rather than *A. cretea* which was proposed in the same paper). Although they are not labelled as such, it is clear the type specimens were collected from Cedar Lake, Manitoba, where all early Canadian amber originated (including the other species which Boesel described).

Types. Holotype, female adult in amber on microscope slide, labelled "Holotype Lasiohelea globosa No 47 "Boesel"", "47", "Atriculicoides globosus (Boesel) Det. A. Borkent" (ROME). Holotype, female adult in amber on microscope slide, labelled "Holotype Lasiohelea cretea No 36 "Boesel"", "36", "Atriculicoides globosus (Boesel) Det. A. Borkent" (ROME); paratype, female adult in amber on microscope slide, labelled "Paratype Lasiohelea cretea No 4 "Boesel"", "4", "Atriculicoides globosus (Boesel) Det. A. Borkent" (ROME).

Material examined. 13 males: CAS 502 (CNCI), CAS 508 (CNCI), CAS 514 (CNCI), CAS 532b (CNCI), CAS 631 (CNCI), CAS 695 (CNCI), CAS 947a (CNCI), CAS 1059 (CNCI), MCZ 9002 (MCZC), MCZ 9003 (MCZC), MCZ 9009 (MCZC), ROM 4 (ROME), 82-15-135 (TMPD); 30 females: CAS 348a, 348b (CNCI), CAS 358 (CNCI), CAS 386 (CNCI), CAS 532a (CNCI), CAS 602 (CNCI), CAS 614b (CNCI), CAS 643 (CNCI), CAS 768 (CNCI), MCZ 6894 (MCZC), MCZ 6907a (MCZC), MCZ 6907b (MCZC), MCZ 9004 (MCZC), MCZ 9005 (MCZC), MCZ 9006 (MCZC), 2 females in MCZ 9007 (MCZC), 10 females in MCZ 9008 (MCZC), ROM 4 (ROME), ROM 33 (ROME), ROM 36 (ROME), ROM 47 (ROME).

Derivation of specific epithet. The name (globular) given by Boesel (1937) certainly refers to rounded basal flagellomeres of the female antenna noted by Boesel (1937) in his description of this species.

Atriculicoides sp.

Diagnosis. Male: unknown. Female: only Cretaceous species of Ceratopogonidae in Canadian amber with eyes broadly contiguous dorsomedially and with palpal segment 3 with a well developed pit.

Description. Female adult. Descriptive statistics: see Tables 9-14.

Head: Eyes bare, broadly contiguous dorsomedially. Vertex with single dorsomedial seta present (Fig. 5D). Antennae with 13 separate flagellomeres, flagellomeres 9-13 more elongate than preceding flagellomeres (Fig. 5E), first flagellomere without sensilla coeloconica. Mouthparts moderately elongate. Clypeus membrane not visible. Labrum with fine teeth. Mandible not visible. Lacinia with five large, retrorse teeth. Palpus with five segments, third segment swollen near midlength, with capitate sensilla arranged in well developed pit (Figs 5G-H). Hypopharynx not visible.

Thorax: Anterior scutal apodemes not visible. Scutum with numerous setae of varying lengths densely scattered over surface, humeral pit not visible. Anapleural suture not visible. Katepisternum without setae.

Wing (Fig. 5C): Macrotrichia present on all membrane, fine microtrichia on all membrane. Alula with macrotrichia. Cells r_1 and r_{2+3} well defined, elongate. M bifurcating distal to r-m.

Legs (Figs 5M-O): Femora, tibiae slender. Legs lacking armature, except foreleg first tarsomere with row of short, thick spines. Fore and midleg trochanter each with pair of thick setae. Midleg tibia without apical spur. Hindleg first tarsomere without thick basal spine, without palisade setae. Foreleg, midleg, hindleg claws equal, with sliver-like inner tooth (Fig. 5I-K). Hindleg claw shorter than fifth tarsomere. With (Fig. 5L) or without (Fig. 5I-K) empodium.

Genitalia: Spermathecae not visible. Sternite 9 continuous medially, without vaginal apodeme. Segment 10 with pair of setae. Cerci of moderate length.

Distribution and bionomics. Specimens of *Atriculicoides* sp. are known from amber collected 5 km south of Grassy Lake, Alberta (n= 1 ♀) and from the shore of Cedar Lake, Manitoba (n= 2 ♀) (Fig. 20B).

Females of *Atriculicoides* sp. likely fed on the blood of unidentified vertebrate hosts (see below under "Evolution of blood-sucking and host selection in Ceratopogonidae").

Taxonomic discussion. Of the material examined, the three females described here as *Atriculicoides* sp. differed significantly from females of *A. globosus*, the only other species of *Atriculicoides* recognized in Canadian amber. The females of *Atriculicoides* sp. had a shorter CR, a palpal segment three with a well developed pit and claws with barely discernable inner teeth (but see "Difficult to interpret character states" regarding the difficulty of accurately characterizing the claws of specimens in amber). One of the three females also had legs with well developed empodia (Fig. 5L).

It is highly likely that these females represent a second species of *Atriculicoides* within the Canadian amber fauna. Considering the morphological variation in the male genitalia described above for *A. globosus* (and hence the likelihood that there is more than one species present), it could easily be that these females are conspecific with some males of what are identified here as *A. globosus*. It is also possible that the females are conspecific with *A. cenomanensis*, known only as a male from French amber, and which also has a well developed palpal pit and a lower CR. Finally, they may represent another species altogether, of which the males are not yet recognized.

Material examined. 3 females: CAS 513 (CNCI), MCZ 6901 (MCZC), ROM 53 (ROME).

Adelohelea Borkent New Genus

Type species: Adelohelea glabra Borkent n.sp., the only included species.

Diagnosis: Male: only genus of Ceratopogonidae with the following combination of character states: with 11 flagellomeres, with two well developed radial cells, and first tarsomere of hindleg without palisade setae. Female: only genus of Ceratopogonidae with the following combination of character states: without basally constricted nipple on apex of flagellomere 13, wing without macrotrichia on membrane, with two well developed, moderately sized, radial cells, first tarsomere of hindleg without palisade setae and claws basally thick and evenly curved.

Description: Very small adults. Eyes bare, narrowly separated dorsomedially. Vertex with single dorsomedial seta present. Male antennae with 11 flagellomeres, some of these at least partially fused, terminal four flagellomeres longer than preceding flagellomeres (Fig. 6C), sensilla coeloconica on first flagellomere not visible, plume moderately well developed, basal foramen of pedicel not visible. Female antenna with 13 separate flagellomeres, flagellomeres 9-13 longer than preceding flagellomeres (Fig. 6E), sensilla coeloconica on first flagellomere not visible. Proboscis moderately short, details of female mouthparts not visible. Palpus with five segments, segment 3 short, semispherical, capitate sensilla arranged in small apical pit. Anterior scutal apodemes not visible. Scutum pruinose, with long scattered setae, humeral pits not visible. Anapleural suture short, extending to posterior margin of anepisternal cleft. Katepisternum without setae. Legs (Fig. 6F-G) slender, unarmed except for apex of each tarsomere of each leg with or without pair of thick setae. Midleg trochanter with pair of thick setae. Midleg tibia without apical spur. Apex of hindleg tibia with 4-5 elongate setae, well developed plumose spur (Fig. 6G). Hindleg first tarsomere straight, without palisade setae, without basal thick spine (Fig. 6G). Female claws equal, short, inner teeth not visible. Male claws apically bifid. Without empodium. Wing with fine microtrichia on membrane, lacking macrotrichia on membrane, alula without macrotrichia, radial veins well developed, M_1, M_2, Cu_1, Cu_2 faint, two well developed radial cells, M bifurcating distal to r-m (Figs 6A,D). Male genitalia of moderate size, with apicolateral process (Fig. 6B). Female genitalia with spermathecae not visible, sternite 9 not visible, cercus of moderate size.

Taxonomic Discussion: The combination of character states given for the male under the diagnosis above are unique within the Ceratopogonidae and serve to distinguish the genus from all others. However, the female may not be separable from some species of *Culicoides*. It is uncertain whether the female of *A. glabra* had sensilla coeloconica on the first flagellomere of the antenna and if future material or study indicates that it is lacking these, this would help to distinguish females of this new taxon.

Derivation of generic name: Adelo (obscure) *helea* (marsh dweller) refers to the enigmatic phylogenetic affinities of this taxon.

Adelohelea glabra Borkent n. sp.

Diagnosis. Male and female: as for genus as this is the only included species.

Description. Male adult. As for genus as this is the only included species. Descriptive statistics: see Tables 3-8.

Genitalia (Fig. 6B): Apicolateral process elongate, tapering gradually to pointed apex, with apical seta. Gonocoxite moderately elongate. Gonostylus thick basally, tapering near apex to somewhat rounded apex, straight. Paramere and aedeagus not visible.

Female adult. As for genus as this is the only included species. Descriptive statistics: see Tables 9-14.

Distribution and bionomics. Adelohelea glabra is known from amber collected 5 km south of Grassy Lake, Alberta (n= 2 ♀) and from the shore of Cedar Lake, Manitoba (n= 1 ♂) (Fig. 20B).

Because the details of the female mouthparts of *A. glabra* were not visible, it is not possible to determine its host (or whether it fed at all).

One female was associated with nine Scatopsidae, two alate homopteran nymphs and what appears to be a hairy seed (only two Scatopsidae in piece with specimen on microscope slide; others in two pieces left in a vial) (79-16-12). Extant Scatopsidae breed in manure or in decaying vegetation or bark. The concentration of Scatopsidae in the amber piece with the female of *Adelohelea glabra* suggests that the original resin was also produced in an area with a significant amount of decaying vegetation or manure.

Taxonomic discussion. Males and females were associated through the shared presence of a wing bare of macrotrichia, two well developed radial cells and the absence of palisade setae on the first tarsomere of the hindleg. The single male from Cedar Lake and the two females from 5 km south of Grassy Lake may indicate that the association is inaccurate. However, evidence presented above ("Botanical source of Canadian amber") suggests that there is no significant difference in the fauna of the two deposits.

Types. Holotype, male adult in amber on microscope slide, labelled "Holotype Adelohelea glabra Borkent, Cedar Lake, Manitoba, August, 1963, J.F. McAlpine & J.E.H. Martin, CAS 530, CNC No. 21663" (CNCI); allotype, female adult in amber on microscope slide, labelled "Allotype Adelohelea glabra Borkent, nr. Medicine Hat [= 5 km s. Grassy Lake], Alta., June, 1974, McAlpine, Cooper & Daze, CAS 1089" (CNCI);

paratype: 1 female, Grassy Lake [= 5 km s. Grassy Lake], Alberta, P 79-16-12 (TMPD).

Derivation of specific epithet. The name *glabra* (hairless, bald) refers to the absence of macrotrichia on the wing membrane of this species.

Heleageron Borkent New Genus

Type species: Heleageron arenatus Borkent n.sp., the only included species.

Diagnosis: Only genus of Ceratopogonidae with the following combination of character states: first flagellomere with sensilla coeloconica, wing without macrotrichia on membrane and only one elongate radial cell, first tarsomere of hindleg without palisade setae, hind tibia with elongate, well developed dorsal setae, female claws basally thick and evenly curved.

Description: Moderately small adults. Eyes bare, narrowly separated dorsomedially. Vertex with single dorsomedial seta present. Male antennae with 13 separate flagellomeres, flagellomeres 11-13 longer than preceding flagellomeres (Figs 6I,J), sensilla coeloconica on first flagellomere not visible, plume well developed, basal foramen of pedicel not visible. Female antenna with 13 separate flagellomeres, flagellomeres 9-13 longer than preceding flagellomeres (Fig. 6N), first flagellomere with sensilla coeloconica (Fig. 6L). Proboscis moderately short, details of female mouthparts not visible. Palpus with five segments, segment 3 moderately long, somewhat ovoid, capitate sensilla and pit not visible. Anterior scutal apodemes not visible. Scutum pruinose, with long scattered setae, humeral pits not visible. Anapleural suture not visible. Katepisternum without setae. Legs slender, unarmed except for apex of each tarsomere of each leg with pair of thick setae. Midleg trochanter with pair of thick setae. Midleg tibia without apical spur. Apex of hindleg tibia with 4-5 elongate setae, well developed plumose spur. Hindleg first tarsomere straight, without palisade setae, without basal thick spine (Fig. 6K). Female claws equal, short, with inner teeth (Figs O-Q). Male claws apically bifid. Empodium not visible. Wing (Fig. 6M) with fine microtrichia on membrane, lacking macrotrichia on membrane, alula without macrotrichia, radial veins well developed, single radial cell, M_1, M_2, Cu_1, Cu_2 faint, M bifurcating distal to r-m. Male genitalia (Fig. 6H) of moderate size, with apicolateral process, with apical or slightly subapical seta. Female genitalia with spermathecae not visible, sternite 9 not visible, cercus of moderate size.

Taxonomic Discussion: The combination of character states given under the diagnosis above are unique within the Ceratopogonidae and serve to distinguish the genus from all others.

Of those extant taxa with a trochanter possessing a strong pair of setae and no palisade setae on the first tarsomere of the hindleg, *Heleageron* shares the presence of an elongate single radial cell with at least two species of *Culicoides* (i.e. *C. uniradialis* Wirth and Blanton (1953), *C. cylindricornis* Wirth and Blanton (1973)). These two species have sensilla coeloconica on other antennal flagellomeres and lack the elongate well developed dorsal setae on the hind tibia of *Heleageron*.

The extant genus *Paradasyhelea* Macfie has reduced or compacted radial veins which extend only a relatively short distance on the wing (Clastrier, 1989; Wirth, 1981; Wirth and Lee, 1959).

Derivation of generic name: Helea (marsh dweller) *geron* (old, aged) refers to the occurrence of this genus in the Cretaceous.

Heleageron arenatus Borkent n. sp.

Diagnosis. Male and female: as for genus as this is the only included species.

Description. Male adult. As for genus as this is the only included species. Descriptive statistics: see Tables 3-8.

Genitalia (Fig. 6H): Apicolateral process elongate, thick, tapering near apex, with apical seta. Gonocoxite somewhat short, thick. Gonostylus more or less straight, thick basally, tapering to somewhat rounded apex. Paramere elongate, slender. Aedeagus not visible.

Female adult. As for genus as this is the only included species. Descriptive statistics: see Tables 9-14.

Distribution and bionomics. Heleageron arenatus is known from amber collected 5 km south of Grassy Lake, Alberta (n= 3 ♂, 2 ♀) and from the shore of Cedar Lake, Manitoba (n= 1 ♂, 2 ♀) (Fig. 20B).

Because the details of the female mouthparts of *H. arenatus* were not visible, it is not possible to determine its host (or whether it fed at all).

One male was associated with a winged male aphid (ROM CL-II-3) and another with an apterous homopteran (82-15-129). One male and two females were associated in a single piece of amber (CAS 1040a, b), indicating that the species was likely at least locally common.

Taxonomic discussion. Males and females were associated through the shared presence of a wing bare of macrotrichia, a single, elongate radial cell and the presence of elongate, well developed dorsal setae on the hind tibiae. In addition, one male and two females of *H. arenatus* were present in a single piece of amber (CAS 1940a, b).

Types. Holotype, male adult in two pieces of amber on microscope slide, labelled "Holotype Heleageron arenatus Borkent, nr. Medicine Hat [= 5 km s. Grassy Lake], Alta., June, 1974, McAlpine, Cooper & Daze, CAS 985, CNC No. 21662" (CNCI); allotype, female adult in amber on microscope slide, labelled "Allotype Heleageron arenatus Borkent, Cedar Lake, Manitoba, F.M.C. Coll., MCZ 9022" (MCZC); paratypes: 1 male,

Grassy Lake [= 5 km s. Grassy Lake], Alberta, 82-15-129 (TMPD); 1 male, Cedar Lake, Manitoba, ROM CL-II-3 (ROME); 1 male, 1 female, near Medicine Hat [= 5 km s. Grassy Lake], Alberta, June, 1974, McAlpine, Cooper & Daze, CAS 1040a (CNCI); 1 female, near Medicine Hat [= 5 km s. Grassy Lake], Alberta, June, 1974, McAlpine, Cooper & Daze, CAS 1040b (CNCI); 1 female, Cedar Lake, Manitoba, MCZ 6897 (MCZC).

Derivation of specific epithet. The name *arenatus* (of sand) refers to the presence of microtrichia on all of the wing membrane.

Culicoides Latreille

Culicoides Latreille, 1809:251. Type species, *Culicoides punctata* Latreille, by monotypy.

Diagnosis. Male: only Ceratopogonidae with sensilla coeloconica on at least the first antennal flagellomere, well developed humeral pits, more or less equal radial cells (or in some, only one cell present), scattered setae on the first tarsomere of the hindleg, hindleg with tibial spur, claws small and equal, and empodium not evident. Female: only Ceratopogonidae with sensilla coeloconica on at least three antennal flagellomere and always including one on the first flagellomere (the Nearctic *C. floridensis* Beck lacks these and the Palaearctic *C. cameroni* Campbell and Pelham-Clinton has only a single sensilla coeloconica, present on the first flagellomere), well developed humeral pits, scattered setae on the first tarsomere of the hindleg, small and equal claws, empodium not evident, and with sternite 9 divided medially (or if continuous medially, with patches of wing pigmentation present).

Description of Adults. As given by Atchley (1967, 1970), Blanton and Wirth (1979), Szadziewski (1988:29), Wirth (1952:165) and papers cited in these. In addition the following features are present: clypeus separated laterally by membrane from head capsule; three anterior scutal apodemes present; anapleural suture short, extending to posterior margin of anepisternal cleft; midleg without a tibial spur.

Distribution and bionomics: Extant species of *Culicoides* are distributed throughout the world with the exception of New Zealand and Patagonia in South America and can be found in a remarkable array of habitats, latitudially from the tropics to the high arctic, altitudinally from tidal flats to high alpine. There are 1,268 species currently recognized (as of 1991). Larvae occur in saline pools, bogs, seeps, streams, rivers, lakes, in decaying vegetative matter, manure, in tree holes and in epiphytes; they are present in virtually any habitat where there is a more or less continuous supply of moisture, oxygen and fine organic material, for at least the time it takes to complete their larval development.

Species of *Culicoides* are relatively common in Canadian amber, making up 35.85% of all Ceratopogonidae (Table 1). This suggests that they formed an important component of the total fauna in the Upper Cretaceous of southern Alberta.

Adults of some extant species are notorious as fierce blood suckers and the females of the vast majority of species have mouthparts that indicate they are active feeders (i.e. serrate mandibles and laciniae with retrorse teeth, toothed labrum and hypopharynx; see below under "Evolution of blood-sucking and host selection in Ceratopogonidae"). Vertebrate hosts include fish (one species, Wirth and Hubert, 1989:305) and reptiles (one species in eastern Nearctic, Jamnback, 1965:106; one species on leatherback turtles in Costa Rica, pers. obs.) but the vast majority feed on either (or both) birds and mammals. Three species of *Culicoides* are known to feed on the abdomens of engorged female mosquitoes under natural conditions (Glukhova, 1982; Wirth and Hubert, 1989:79).

The correlation of host type and modifications of the female mouthparts is discussed below under "Evolution of blood-sucking and host selection in Ceratopogonidae". Interpretation of the female mouthparts of the fossil species indicates that all of these (for which females are known) fed on vertebrates and that two, and possibly three, species fed on large dinosaurs.

Taxonomic Discussion: There are now 37 recognized fossil species of *Culicoides* and these are recorded from the Upper Cretaceous to the Miocene (Table 18). Eleven of these are from Cretaceous ambers; seven from Canadian, four from Taimyr Peninsula and one from New Jersey. Of the seven species recognized in Canadian amber, one is considered as conspecific with a species in Taimyr amber and four are described here as new.

Based on a single holotype female, Remm (1976) suggested that *C. kaluginae* Remm has broadly separated eyes (his Fig. 2a) and this was also reported as such by Glukhova (1989:101). If this character state is accurately described, this species likely does not belong within *Culicoides*, but is closely related to *Leptoconops*. This character is discussed further below under synapomorphy 13 in "Phylogenetic relationships of extant Ceratopogonidae". However, it is possible that this feature is an artifact of observation. One female specimen of *C. tyrrelli* (82-15-22) which was examined during the course of this study had what initially appeared to be widely spaced eyes. The position of the body was very similar to that portrayed by Remm (1976: fig. 2a). However, upon closer examination under 312X, the eyes could be seen to be only narrowly separated. It would be very informative to re-examine the holotype of *C. kaluginae* to determine if the eyes are really broadly separated.

Of the Canadian Cretaceous material identified as *Culicoides,* the association between the males and females of *C. filipalpis* seems well established (see below under that species) and both males and females of this species were the most common *Culicoides* in the amber (Table 1). The associations made between the sexes of the remaining material, which included six male and three female forms, was extremely tenuous. Of the six named species, the male and female of *C. tyrrelli* had the longest mouthparts (Tables 5, 10).

Of the remaining five male and two female forms, the males and females of *C. bullus* had relatively more of the wing membrane covered with macrotrichia. Finally of the remaining four male and single female forms, the males and females of *C. canadensis* were associated on the basis of the lower relative lengths of palpal segments three and four (Tables 6, 11). This process of elimination left three species (*C. agamus, C. obuncus*, and *C. annosus*) without any associated females. However, the males of these three species have somewhat longer wings than males of *C. canadensis* (Table 7) and the longer wings of at least some female *C. canadensis* (Table 12) may indicate they actually belong to any of these unassociated species here recognized solely as males.

It was clear that I did not have enough material to be confident of the association of males and females of these latter six species. Certainly the future study of longer series of specimens would be invaluable but it should be noted that, even so, it may be impossible to confidently associate males and females of these morphologically similar species.

Similarly, some of the differences recorded as distinguishing some species of *Culicoides* from Canadian amber from those in Taimyr amber may be inaccurate or artifacts. It is imperative that future comparisons of specimens from these two localities be made to ensure that some of the species are actually distinct. Some striking similarities and suspect differences are discussed below under the appropriate species.

Key to Cretaceous Fossil *Culicoides*

The male of *C. kaluginae* and the female of *C. casei* are unknown. The distinctions between the species known only from Taimyr amber are based on the descriptions given by Remm (1976) and Glukhova (1989).

1. First tarsomere of the midleg with small spines (only known as female) . . . . . .*kaluginae* Remm (Taimyr amber, Upper Cretaceous)
1' First tarsomere of the midleg with only slender setae 2
2. Male 3
2' Female 12

3. Wing with only a few macrotrichia on membrane, near the very apex of the wing . . . . *casei* Grogan and Szadziewski (New Jersey amber, Upper Cretaceous)
3' Wing with many macrotrichia (i.e. more than 40) on at least the apical half of the wing membrane (Figs 12A,I). 4
4. Gonostylus divided apically or with inner, subapical lobe (Figs 7D,F, 9D-E) 5
4' Gonostylus tapering to pointed apex (Figs 10A,11A,H,12A,I) 6
5. Gonostylus with tricuspate apex (Figs 7D,F) . . . . *filipalpis* Remm (Canadian and Taimyr amber, Upper Cretaceous)
5. Gonostylus with inner subapical lobe or appearing to have bifurcate apex (Figs 9D-E) . . . . *tyrrelli* (Boesel) (Canadian amber, Upper Cretaceous)
6. Gonostylus with a thick base, gradually tapering to apex (Figs 10E-F,11L,12C,M-N) 7
6' Gonostylus slender, of nearly even diameter for most of length (Figs 11C) 11
7. Wing with macrotrichia restricted to area distal to fork of M and Cu (Figs 12A,I). 8
7' Wing with macrotrichia present basal of fork of M and Cu (Figs 10A,11H) 10
8. Apicolateral process gradually tapering to pointed apex. . . . . *sphenostylus* Remm (Taimyr amber, Upper Cretaceous)
8' Apicolateral process cylindrical for most of its length, tapering only near very apex (Figs 12C,M) 9
9. Gonostylus chelate, with mediobasal lobe (Fig. 12C) . . . . *obuncus* n.sp. (Canadian amber, Upper Cretaceous)
9' Gonostylus gradually tapering from thick base (Figs M-N). . . . . *annosus* n.sp. (Canadian amber, Upper Cretaceous)
10. Apicolateral process thick, broad, tapering near apex (Figs 10E-F) . . . . . . *canadensis* (Boesel) (Canadian amber, Upper Cretaceous)
10' Apicolateral process tapering gradually to narrow apex (Fig. 11L) . . . . *agamus* n.sp. (Canadian amber, Upper Cretaceous)
11. Apicolateral processes of moderate size, tapering to apex; gonostylus very slender for entire length . . . . *succineus* Szadziewski (Taimyr amber, Upper Cretaceous)
11' Apicolateral processes very small, knob-like; gonostylus thick for entire length (Fig. 11C) . . . . *bullus* n.sp. (Canadian amber, Upper Cretaceous)
12. Mouthparts long (Figs 8G,9L), mouthpart length/length of foreleg fifth tarsomere > 3.5 13
12' Mouthparts of moderate length (Fig. 10G), mouthpart length/length of foreleg fifth tarsomere < 3.6 14

13. Palpal segment three long, slender and cylindrical (Figs 8G,K). . . .
filipalpis Remm (Taimyr and Canadian amber, Upper Cretaceous)
13' Palpal segment three long, swollen at midlength to subapically, capitate sensilla in shallow, broad pits (Fig. 9L) . . . . *tyrrelli* (Boesel) (Canadian amber, Upper Cretaceous)
14. Wing with macrotrichia present over entire membrane (Fig. 11D)
. . . . *succineus* Remm (Taimyr amber, Upper Cretaceous)
. . . . *bullus* n.sp. (Canadian amber, Upper Cretaceous)
14' Wing with fewer macrotrichia on membrane, absent from basal portion of cell cua_2 (Fig. 10H) 15
15. Wing with some macrotrichia in cell cua_2 . . . .
. . . . . . *canadensis* (Boesel) (Canadian amber, Upper Cretaceous)
15' Wing without macrotrichia in cell cua_2 . . . .
. . . . *sphenostylus* Remm (Taimyr amber, Upper Cretaceous)

Derivation of generic name: The name *Culicoides* (*Culex*, mosquito, -*oides*, resembling) given by Latreille (1809), likely refers to the perceived similarities of the included species to the Culicidae.

Culicoides filipalpis Remm

Culicoides filipalpis Remm, 1976:111. Yantardakh area, Taimyr Peninsula, Russia. Upper Cretaceous.

Diagnosis. Male: only species of Cretaceous Ceratopogonidae with the apex of the gonostylus tricuspate. Female: only species of Cretaceous Ceratopogonidae with a markedly elongate palpus in which segments 2 and 3 are long, slender and cylindrical.

Description. Male adult. Descriptive statistics: see Tables 3-8.

Head: Eyes bare, narrowly separated dorsomedially. Vertex with single dorsomedial seta present. Antenna with well developed plume, basal foramen of pedicel not visible, 13 separate flagellomeres (in some specimens, flagellomeres 2-10 appear at least partially fused), flagellomeres 11-13 and possibly 10 more elongate than preceding flagellomeres (Fig. 7B), first flagellomere with sensilla coeloconica. Mouthparts elongate. Palpus with five segments, second and third segments elongate, cylindrical, third segment with capitate sensilla arranged on surface or in shallow depression (Fig. 7C).

Thorax: With three anterior scutal apodemes (Fig. 7E). Scutum pruinose, with scattered elongate setae, humeral pit well developed. Anapleural suture not visible. Katepisternum without setae.

Wing (Fig. 7A): Macrotrichia present on all membrane, fine microtrichia present on all membrane. Alula with macrotrichia. Cells r_1 and r_{2+3} well defined, short. M bifurcating distal to r-m.

Legs: Femora, tibiae slender. Legs lacking armature, except for apex of each tarsomere of each leg with pair of thick setae. Fore and midleg trochanter each with pair of thick setae. Midleg tibia without apical spur. Hindleg first tarsomere without thick basal spine, without palisade setae. Claws apically bifid. Without empodium.

Genitalia (Figs 7D,F-H): Apicolateral process elongate, tapering or cylindrical, with apical seta. Gonocoxite short, thick. Gonostylus thick basally, tapering to tricuspate apex. Paramere well developed, separate, tapering to sinuous, elongate apex. Aedeagus forming single median structure with short, tapering posterior extension.

Female adult. Descriptive statistics: see Tables 9-14.

Head: Eyes bare, narrowly separated dorsomedially. Vertex with single dorsomedial seta present. Antennae with 13 separate flagellomeres, flagellomeres 9-13 more elongate than preceding flagellomeres (Fig. 8B), first flagellomere with sensilla coeloconica (Figs 8D-E). Mouthparts markedly elongate (Fig. 8G). Clypeus separated laterally by membrane from head capsule. Labrum with large teeth (Fig. 8I). Mandible with fine teeth (Fig. 8H). Lacinia with very fine teeth (Fig. 8J). Palpus with five segments, third segment markedly elongate, nearly cylindrical, with capitate sensilla arranged on surface or in shallow depression (Fig. 8K). Hypopharynx not visible.

Thorax: With three anterior scutal apodemes. Scutum pruinose, with scattered elongate setae, humeral pit well developed. Anapleural suture short, extending to posterior margin of anepisternal cleft. Katepisternum without setae.

Wing (Fig. 8A): Macrotrichia present on all membrane, fine microtrichia present on all membrane. Alula with macrotrichia. Cells r_1 and r_{2+3} well defined, short to moderately elongate. M bifurcating distal to r-m.

Legs: Femora, tibiae slender. Legs lacking armature, except for apex of each tarsomere of each leg with pair of thick setae. Fore and midleg trochanter each with pair of thick setae. Midleg tibia without apical spur. Hindleg first tarsomere without thick basal spine, without palisade setae (Fig. 8F). Foreleg, midleg, hindleg claws equal, simple, with very slender inner tooth arising from near base of each talon (not visible in some) (Fig. 8C). Hindleg claw shorter than fifth tarsomere. Without empodium.

Genitalia (Fig. 8L): Spermathecae not visible. Sternite 9 divided medially, without vaginal apodeme. Setae on segment 10 not visible. Cerci of moderate length.

Distribution and bionomics. Culicoides filipalpis is known from amber collected from the Taimyr Peninsula in Russia (n= 2 ♀, not seen) and within Canada, from 5 km south of Grassy Lake, Alberta (n= 2 ♂, 9 ♀) and from the shore of Cedar Lake, Manitoba (n= 10 ♂, 3 ♀) (Fig. 20B).

Female *C. filipalpis* likely fed on the blood of vertebrate hosts and possibly on that of dinosaurs (see below under "Evolution of blood-sucking and host selection in Ceratopogonidae").

There are a few associated organisms present in the amber containing members of *C. filipalpis*: a male specimen was associated with an Hymenoptera (MCZ 9010) and of females, one was associated with a larval mite which may be a Parasitengona (CAS 777) and another was associated with a male *Protoculicoides depressus*.

Two pieces of amber held more than one individual of *C. filipalpis:* amber piece MCZ 9011 contained two males and 82-15-40 held three females. These observations indicate that *C. filipalpis* was likely common in at least certain habitats and, perhaps, that the males aggregated to form swarms.

Taxonomic discussion. The males and females of *C. filipalpis* were associated through the shared presence of distinctively long mouthparts in which the third palpal segment was strikingly elongate and cylindrical.

C. filipalpis was originally recognized on the basis of two females from Taimyr amber (Remm, 1976). The presence of a significant medial gap between the eyes described by Remm (1976) is here considered to be an artifact (see discussion above under 'Taxonomic Discussion' of *Culicoides*).

Because I was unable to examine any material from Taimyr amber, it will certainly be desirable for future work to make direct comparisons between Old and New World material of this species to confirm that they are conspecific.

Types. Not seen.

Material examined. 12 males: CAS 504 (CNCI), CAS 515 (CNCI), CAS 519 (CNCI), CAS 986 (CNCI), CAS 1246 (CNCI), MCZ 9010 (MCZC), 2 males in MCZ 9011 (MCZC), MCZ 9012 (MCZC), MCZ 9013 (MCZC), MCZ 9014 (MCZC), ROM 55 (ROME); 12 females: CAS 416 (CNCI), CAS 492 (CNCI), CAS 667 (CNCI), CAS 777 (CNCI), CAS 948 (CNCI), CAS 1015 (CNCI), MCZ 9015 (MCZC), MCZ 9016 (MCZC), MCZ 9017 (MCZC), 3 females in 82-15-40 (TMPD).

Derivation of specific epithet. The name *filipalpis* (*filum* -thread, *palpus* - palpus) given by Remm (1976) undoubtedly refers to the elongate palpus of the female of this species.

Culicoides tyrrelli (Boesel) NEW COMBINATION

Dasyhelea tyrrelli Boesel, 1937:48. Cedar Lake, Manitoba. Upper Cretaceous.

Diagnosis. Male: only Cretaceous species of Ceratopogonidae in which the gonostylus bears a subapical, lobe projecting dorsomedially (gonostylus appearing apically bifurcate from some angles). Female: only Cretaceous species of Ceratopogonidae with markedly elongate mouthparts and with a third palpal segment swollen at midlength or subapically and with a broad, shallow pit.

Description. Male adult. Descriptive statistics: see Tables 3-8.

Head: Eyes bare, narrowly separated dorsomedially. Vertex with single dorsomedial seta present. Antenna with well developed plume, basal foramen of pedicel not visible, 13 flagellomeres with either separate or flagellomeres 2-10 at least partially fused, flagellomeres 11-13 more elongate than preceding flagellomeres (Fig. 9C), first flagellomere with sensilla coeloconica. Mouthparts elongate. Palpus with five segments, third segment elongate, somewhat swollen near midlength, with capitate sensilla in shallow pit (Fig. 9B).

Thorax: Anterior scutal apodemes not visible. Scutum pruinose, with scattered elongate setae, humeral pit not visible. Anapleural suture not visible. Katepisternum without setae.

Wing (Fig. 9A): Macrotrichia present on membrane on about apical half of wing, fine microtrichia present on all membrane. Alula with macrotrichia. Cells r_1 and r_{2+3} well defined, short. M bifurcating distal to r-m.

Legs: Femora, tibiae slender. Legs lacking armature. Fore and midleg trochanter each with pair of thick setae. Midleg tibia without apical spur. Hindleg first tarsomere without thick basal spine, without palisade setae. Claws apically bifid. Without empodium.

Genitalia (Figs 9D-E): Apicolateral process elongate, slender, cylindrical, with apical seta. Gonocoxite short, thick. Gonostylus with inner, subapical lobe, gonostylus appearing apically bifurcate from some angles. Paramere not visible. Aedeagus not visible.

Female adult. Descriptive statistics: see Tables 9-14.

Head: Eyes bare, narrowly separated dorsomedially. Vertex with single dorsomedial seta present. Antennae with 13 separate flagellomeres, flagellomeres 9-13 more elongate than preceding flagellomeres (Fig. 9M), first flagellomere with sensilla coeloconica (Figs 9N-O). Mouthparts elongate (Fig. 9L). Clypeus separated laterally by membrane from head capsule. Labrum with fine teeth (Fig. 9P). Mandible with fine teeth. Lacinia with numerous very fine teeth (Fig. 9P). Palpus with five segments, third segment moderately elongate, swollen at midlength to near apex, with capitate sensilla in well developed pit (Fig. 9L). Hypopharynx with teeth.

Thorax: Anterior scutal apodemes not visible. Scutum pruinose, with scattered elongate setae, humeral pit not visible. Anapleural suture not visible. Katepisternal setae not visible.

Wing (Fig. 9F): Macrotrichia present on most of membrane but sparse or absent on basal portion of wing, fine microtrichia present on all membrane. Alula with macrotrichia. Cells r_1 and r_{2+3} well defined, short. M bifurcating distal to r-m.

Legs (Figs 9I-K): Femora, tibiae slender. Legs lacking armature. Fore and midleg trochanter each with pair of thick setae. Midleg tibia without apical spur. Hindleg first tarsomere without thick basal spine, without palisade setae. Foreleg, midleg, hindleg claws equal, simple, with very slender inner tooth arising from near base of each talon. Hindleg claw shorter than fifth tarsomere. Without empodium.

Genitalia: Spermathecae not visible. Sternite 9 not clearly visible but in one specimen (82-15-22) it is likely divided, without vaginal apodeme. Segment 10 not visible. Cerci of moderate length.

Distribution and bionomics. C. tyrrelli is known from amber collected 5 km south of Grassy Lake, Alberta (n= 1 ♂, 4 ♀) and from the shore of Cedar Lake, Manitoba (n= 1 ♂, 3 ♀) (Fig. 20B).

Female *C. tyrrelli* likely fed on the blood of large dinosaurs (see below under "Evolution of blood-sucking and host selection in Ceratopogonidae").

There were some associated organisms in amber containing specimens of *C. tyrrelli*. With one of the males (CAS 287) there was a Braconidae of the subfamily Hormiinae (possibly *Hormius* Nees, a gregarious ectoparasite of Lepidoptera), a Sciadoceridae (Diptera) and a trichome (a brittle star-shaped structure found on the surface of many plants) with five arms each with a length of about 0.15 mm present. In pieces holding females, one also had a trichome with eight arms each with a length of about 0.15 mm (CAS 565), one held two Hymenoptera: one unidentifiable beyond order level, the other a *Serphites* Brues (Serphitidae) (ROM 48) and one was associated with a male *Atriculicoides globosus* (MCZ 9009).

Taxonomic discussion. Males and females were associated through the shared presence of the following combination of character states: sensilla coeloconica on the first flagellomeres, wings with only a moderate amount of macrotrichia (sparse or absent on basal portion of the wing), elongate mouthparts, and palpal segment 3 of moderate length.

Females *C. tyrrelli* may actually include two species. Specimen CAS 565 (CNCI) was much smaller than other members with a wing length of 0.56 mm; the next smallest specimen had a wing length of 0.81 mm.

The amber piece holding the holotype of *C. tyrrelli* was further cut and polished for this study and the specimen examined from a number of angles.

One of the females of *C. tyrrelli* (ROM 48) had been previously identified as a member of *Lasiohelea* Kieffer by Boesel.

Boesel (1937) indicated that the wing of the holotype likely had more macrotrichia than is presently evident on the wing, as there were loose hairs present in the amber. The loose hairs could not be confirmed and it is likely that Boesel was mislead by fine contamination in the amber dorsal to the wings when the amber piece was larger than it is now. The extent of macrotrichia on the wing as shown (Fig. 9A; Boesel, 1937, fig. 11A) is likely an accurate depiction of the condition in this species.

Types. Holotype, male adult in amber on microscope slide, labelled "Holotype Dasyhelea tyrrelli No. 18 "Boesel"", "Cedar Lake, Manitoba, Culicoides tyrrelli, Det. A. Borkent, ROM 28" (ROME).

Material examined. 2 males: ROM 28 (ROME), CAS 287 (separate piece with associated species stored in small box) (CNCI); 7 females: CAS 565 (CNCI), CAS 855 (CNCI), CAS 944 (CNCI), MCZ 9009 (MCZC), MCZ 9018 (MCZC), ROM 48 (ROME), 82-15-22a (TMPD).

Derivation of specific epithet. The name *tyrrelli* was given by Boesel (1937) in honor of J.B. Tyrrell who first studied the amber deposits on Cedar Lake, Manitoba in 1890 and 1891.

Culicoides canadensis (Boesel) NEW COMBINATION

Atrichopogon canadensis Boesel, 1937:47. Cedar Lake, Manitoba. Upper Cretaceous.

Diagnosis. Male: only Cretaceous species of Ceratopogonidae with extensive macrotrichia on at least the distal half of the wing, with a gonostylus with a broad base tapering to a single pointed apex and with a broad, thick apicolateral process. Female: only Cretaceous species of *Culicoides* with macrotrichia absent from membrane on very basal portion of wing but present in cell cu.

Description. Male adult. Descriptive statistics: see Tables 3-8.

Head: Eyes bare, narrowly separated dorsomedially with single dorsomedial seta present. Antenna with well developed plume, basal foramen of pedicel not visible, 13 separate flagellomeres, flagellomeres 11-13 more elongate than preceding flagellomeres (Fig. 10B), first flagellomere with sensilla coeloconica (Fig. 10C). Mouthparts moderately elongate. Palpus with five segments, third segment moderately short, with capitate sensilla arranged on surface or in shallow depression (Fig. 10D).

Thorax: Scutal apodemes not visible. Scutum pruinose, with scattered elongate setae, humeral pit not visible. Anapleural suture not visible. Katepisternum without setae.

Wing (Fig. 10A): Macrotrichia present on membrane on about apical half of wing, fine microtrichia present on all membrane. Alula with macrotrichia. Cells r_1 and r_{2+3} well defined, short. M bifurcating distal to r-m.

Legs: Femora, tibiae slender. Legs lacking armature. Fore and midleg trochanter each with pair of thick setae. Midleg tibia without apical spur. Hindleg first tarsomere without thick basal spine or palisade setae. Claws apically bifid. Without empodium.

Genitalia (Figs 10E-F): Apicolateral process elongate, thick, tapering near apex, with apical seta. Gonocoxite moderately elongate. Gonostylus thick basally, tapering to pointed apex. Paramere well developed, separate, tapering to sinuous, elongate apex. Aedeagus with well developed basal apodemes, with tapering posterior extension.

Female adult. Descriptive statistics: see Tables 9-14.

Head: Eyes bare, narrowly separated dorsomedially. Vertex with single dorsomedial seta present (Fig. 10G). Antennae with 13 separate flagellomeres, flagellomeres 9-13 more elongate than preceding flagellomeres (Fig. 10K), sensilla coeloconica on first flagellomere not visible. Mouthparts moderately elongate (Fig. 10G). Clypeus separated laterally by membrane from head capsule. Labrum and/or mandible with fine teeth (identification of mouthpart not certain) (Fig. 10I). Lacinia with numerous fine teeth (Fig. 10I). Palpus with five segments, third segment moderately elongate, with 10 (seen in one specimen) capitate sensilla arranged on surface or in shallow depression (Fig. 10J). Hypopharynx not visible.

Thorax: With three anterior scutal apodemes. Scutum pruinose, with scattered elongate setae, humeral pit well developed. Anapleural suture short, extending to posterior margin of anepisternal cleft. Katepisternum without setae.

Wing (Fig. 10H): Macrotrichia present on membrane on about apical half of wing, fine microtrichia present on all membrane. Alula with macrotrichia. Cells r_1 and r_{2+3} well defined, short. M bifurcating distal to r-m.

Legs (Fig. 10O-Q): Femora, tibiae slender. Legs lacking armature. Fore and midleg trochanter each with pair of thick setae. Midleg tibia without apical spur. Hindleg first tarsomere without thick basal spine, without palisade setae. Foreleg, midleg, hindleg claws equal, simple, with very slender inner tooth arising from near base of each talon (Figs 10L-N). Hindleg claw shorter than fifth tarsomere (Fig. 10N). Without empodium.

Genitalia: Spermathecae not visible. Sternite 9 not visible medially, without vaginal apodeme. Segment 10 setae not visible. Cerci of moderate length.

Distribution and bionomics. C. canadensis is known from amber collected 5 km south of Grassy Lake, Alberta (n= 4 ♂, 4 ♀) and from the shore of Cedar Lake, Manitoba (n= 1 ♂, 2 ♀) (Fig. 20B).

Female *C. canadensis* likely fed on the blood of large dinosaurs (see below under "Evolution of blood-sucking and host selection in Ceratopogonidae"). One female had a somewhat enlarged abdomen, suggesting the possibility that this individual had taken a blood-meal (or possibly nectar) shortly before preservation in resin.

There were a few associated organisms in the amber containing specimens of *C. canadensis*. One male (CAS 902) was associated with an alate nymph of the aphid *Mesozoicaphis canadensis* Heie and the females had the following also present in their amber: legs of a Scatopsidae (identification uncertain) (ROM 35); legs of a Phoridae (identification uncertain) (CAS 516); mite (not presently identifiable beyond this level) (CAS 721); a female Chironomidae and several trichomes (brittle star-shaped structures found on the surface of many plants) with 7-16 arms, each with a length of about 0.1-0.2 mm (CAS 636).

The presence of an associated chironomid with one of the females suggests the proximity of aquatic habitats.

Taxonomic discussion. Males and females were associated through the shared presence of only moderately elongate mouthparts, the absence of macrotrichia on the base of the wing, and the relative lengths of palpal segments three and four. The tenuous nature of this association is discussed further above under 'Taxonomic Discussion' of the genus *Culicoides*. It is likely that the females of *C. canadensis* include two or more species but no objective means could be found to distinguish between members included in the series.

C. canadensis is very similar to *C. sphenostylus* Remm from Taimyr amber and is mainly distinguished on the basis of the more extensive macrotrichia on the wing base of *C. canadensis*. In addition, there may be differences in the shape of the apicolateral processes of the males of the two species. The drawings of *C. sphenostylus* (Remm, 1976: fig. 3f, 3g) indicates that the apicolateral processes of that species taper rapidly to sharply pointed apices.

As a new combination, *Culicoides canadensis* (Boesel) is a senior homonym of *Culicoides canadensis* Wirth & Blanton (1969:211) which is here renamed *Culicoides paraimpunctatus* NEW NAME.

Types. Holotype, female adult in amber on microscope slide, labelled "Holotype Atrichopogon canadensis No. 35 "Boesel"", "35", "Culicoides canadensis, Det. A. Borkent" (ROME).

Material examined. 5 males: CAS 830 (CNCI), CAS 831 (CNCI), CAS 902 (CNCI), CAS 1027 (CNCI), MCZ 9019 (MCZC); 6 females: CAS 507 (CNCI), CAS 516 (CNCI), CAS 636 (CNCI), CAS 721 (CNCI), CAS 1154 (CNCI), ROM 35 (ROME).

Derivation of specific epithet. The name *canadensis* given by Boesel refers to the type locality in Canada.

Culicoides bullus Borkent n. sp.

Diagnosis. Male: only Cretaceous species of *Culicoides* with a more or less straight gonostylus of nearly constant but thick diameter and with small, bud-like apicolateral processes. Female: only Cretaceous species of *Culicoides* with only moderately elongate mouthparts (mouthpart length/length of foreleg fifth tarsomere < 3.6) and with macrotrichia covering all wing membrane (not distinguishable from *C. succineus* Remm from Taimyr amber).

Description. Male adult. Descriptive statistics: see Tables 3-8.

Head: Eyes bare, narrowly separated dorsomedially. Vertex with single dorsomedial seta not visible. Antenna with well developed plume, basal foramen of pedicel not visible, 13 flagellomeres (separation or fusion not visible), flagellomeres 11-13 more elongate than preceding flagellomeres (Fig. 11B), sensilla coeloconica on first flagellomere not visible. Mouthparts moderately elongate. Palpus with five segments, third segment elongate, somewhat swollen at midlength, with capitate sensilla not visible.

Thorax: Scutal apodemes not visible. Scutum with at least some elongate setae (details not visible), humeral pit not visible. Anapleural suture not visible. Katepisternum without setae.

Wing (Fig. 11A): Macrotrichia present on all membrane, fine microtrichia present on all membrane. Alula with macrotrichia. Cells r_1 and r_{2+3} well defined, short. M bifurcating distal to r-m.

Legs: Femora, tibiae slender. Legs lacking armature. Fore and midleg trochanter each with pair of thick setae. Midleg tibia without apical spur. Hindleg first tarsomere without thick basal spine, without palisade setae. Claws apically bifid. Without empodium.

Genitalia (Fig. 11C): Apicolateral process short, bud-like, with apical seta. Gonocoxite moderately elongate. Gonostylus of more or less uniform diameter, apex bent outwardly. Paramere not visible. Aedeagus forming single broad median structure with well developed basal apodemes and forked posterior extension.

Female adult. Descriptive statistics: see Tables 9-14.

Head: Eyes bare, narrowly separated dorsomedially. Vertex with single dorsomedial seta not visible. Antennae with 13 separate flagellomeres, flagellomeres 9-13 more elongate than preceding flagellomeres (Fig. 11G), first flagellomere with sensilla coeloconica. Mouthparts moderately elongate. Separation of clypeus from head capsule not visible. Labrum not visible. Mandible with fine teeth (Fig. 11F). Lacinia with more than 15 fine teeth (Fig. 11F). Palpus with five segments, third segment somewhat ovoid, with about seven (difficult to see accurately) capitate sensilla arranged in well developed pit (Fig. 11E). Hypopharynx not visible.

Thorax: Scutal apodemes not visible. Scutum pruinose, with scattered elongate setae, humeral pit not visible. Anapleural suture not visible. Katepisternum without setae.

Wing (Fig. 11D): Macrotrichia present on all membrane, fine microtrichia present on all membrane. Alula with macrotrichia. Cells r_1 and r_{2+3} well defined, short. M bifurcating distal to r-m.

Legs: Femora, tibiae slender. Legs lacking armature. Fore and midleg trochanter each with pair of thick setae. Midleg tibia without apical spur. Hindleg first tarsomere without thick basal spine, without palisade setae. Foreleg, midleg, hindleg claws equal, simple, with very slender inner tooth arising from near base of each talon. Hindleg claw shorter than fifth tarsomere. Without empodium.

Genitalia: Spermathecae not visible. Sternite 9 not visible. Segment 10 setae not visible. Cerci of moderate length.

Distribution and bionomics. C. bullus is known from amber collected 5 km south of Grassy Lake, Alberta (n= 1 ♂, 2 ♀) (Fig. 20B).

Female *C. bullus* likely fed on the blood of unidentified vertebrate hosts (see below under "Evolution of blood-sucking and host selection in Ceratopogonidae").

Taxonomic discussion. Males and females were associated through the shared presence of relatively short mouthparts (compared to *C. filipalpis, C. tyrrelli*) and extensive macrotrichia on the wing membrane.

The females of *C. bullus* cannot be distinguished from those of *C. succineus* from Taimyr amber (Remm, 1976). The differences in the CR (*C. bullus* = 0.62-0.64; *C. succineus* = 0.56 (taken from Remm, 1976: fig. 3h)) may be significant but this needs to be tested with actual examination of the Taimyr specimens.

Types. Holotype, male adult in amber on microscope slide, labelled "Holotype Culicoides bullus Borkent, nr. Medicine Hat [= 5 km s. Grassy Lake], Alta, 1973, P. Boston, CAS 791, CNC No. 21661" (CNCI); allotype, female adult in amber on microscope slide, labelled "Allotype Culicoides bullus Borkent, Grassy Lake [= 5 km s. Grassy Lake], Alta, Tyrrell 82-15-20" (TMPD); paratype: 1 female, Grassy Lake [= 5 km s. Grassy Lake], Alberta, 82-15-90 (TMPD).

Derivation of specific epithet. The name *bullus* (knob) refers to the small apicolateral processes of the male genitalia.

Culicoides agamus Borkent n. sp.

Diagnosis. Male: only Cretaceous species of Ceratopogonidae with a wing covered with macrotrichia, moderately long apicolateral processes which gradually taper to a narrow apex and with a gonostylus with a broad base tapering to narrow apex. Female: unknown.

Description. Male adult. Descriptive statistics: see Tables 3-8.

Head: Eye pubescence not visible. Dorsomedial separation of eyes not visible. Vertex with single dorsomedial seta present. Antenna with well developed plume, basal foramen of pedicel not visible, 13 separate flagellomeres (or with 2-6 appearing fused), flagellomeres 11-13 more elongate than preceding flagellomeres (Fig. 11I), first flagellomere with sensilla coeloconica (Fig. 11J). Mouthparts moderately elongate. Palpus with five segments, third segment moderately elongate, somewhat swollen at midlength, with capitate sensilla not visible (Fig. 11K).

Thorax: With three anterior scutal apodemes. Scutum pruinose, with scattered elongate setae, humeral pit not visible. Anapleural suture not visible. Katepisternum without setae.

Wing (Fig. 11H, very curved in amber): Macrotrichia present on all membrane but concentrated along wing veins, fine microtrichia present on all membrane. Alula with macrotrichia. Cells r_1 and r_{2+3} well defined, moderately short. M bifurcating distal to r-m.

Legs: Femora, tibiae slender. Legs lacking armature or in one specimen (MCZ 9021) with apex of each tarsomere of each leg with pair of thick setae. Pair of thick setae on fore and midleg trochanter not visible. Midleg tibia without apical spur. Hindleg first tarsomere without thick basal spine, without palisade setae. Apex of claws not clearly visible. Without empodium.

Genitalia (Fig. 11L): Apicolateral process elongate, tapering gradually from base to pointed apex, with apical seta. Gonocoxite moderately short. Gonostylus thick basally, tapering to pointed apex. Aedeagus and parameres not clearly visible but paired, apically rounded structures present which could be either.

Distribution and bionomics. C. agamus is known from amber collected from the shore of Cedar Lake, Manitoba (n= 3 ♂) (Fig. 20B).

Because the female of *C. agamus* is unknown, it is not possible to determine its host (or whether it fed at all).

Two males were associated in one piece of amber (MCZ 9020), but appear to be in different states of preservation and possibly in different planes in the amber (indicated by differences in 'flow' lines within the amber). This suggests that the two males were caught by the original resin at different times. Two males of the same species in the amber may, however, indicate that the species was locally common.

Taxonomic discussion. There is some question as to whether the male in amber piece MCZ 9021 is conspecific with the two in piece MCZ 9020. The single male has its genitalia in an awkward position for accurate viewing and may be significantly larger.

Types. Holotype, the more translucent male adult with both antennae present, in amber piece MCZ 9020 on microscope slide, labelled "Holotype Culicoides agamus Borkent, Cedar Lake, Manitoba, F.M.C. Coll., MCZ 9020" (MCZC); paratypes: 2 males from type locality, MCZ 9020 (in same piece as holotype) (MCZC), MCZ 9021 (MCZC).

Derivation of specific epithet. The name *agamus* (bachelor, unmarried) refers to the fact that only males are presently known for this species.

Culicoides obuncus Borkent n. sp.

Diagnosis. Male: only Cretaceous species of Ceratopogonidae with a wing bearing only a few macrotrichia on apical half of the wing, and with a chelate-shaped gonostylus (i.e. with a broad inner basal lobe). Female: unknown.

Description. Male adult. Descriptive statistics: see Tables 3-8.

Head: Eyes bare, dorsomedial separation not visible. Vertex with single dorsomedial seta present. Antenna with well developed plume, basal foramen of pedicel not visible, 13 separate flagellomeres, flagellomeres 11-13 more elongate than preceding flagellomeres (Fig. 12B), first flagellomere with sensilla coeloconica (Fig. 12D). Mouthparts moderately elongate. Palpus with five segments, third segment moderately elongate, somewhat swollen at midlength, with a few capitate sensilla arranged in a shallow pit (Fig. 12E).

Thorax: Scutal apodemes not visible. Scutum pruinose, with scattered elongate setae, humeral pit not visible. Anapleural suture not visible. Katepisternum without setae.

Wing (Fig. 12A): Macrotrichia present on some or all of terminal wing veins, with some scattered on membrane at apex of wing, fine microtrichia present on all membrane. Alula with macrotrichia. Cells r_1 and r_{2+3} well defined, short. M bifurcating distal to r-m.

Legs: Femora, tibiae slender. Legs lacking armature or in some specimens with apex of each tarsomere of at least some legs with pair of thick setae. Fore and midleg trochanter with pair of thick setae. Midleg tibia without apical spur. Hindleg first tarsomere without thick basal spine, without palisade setae. Claws apically bifid. Without empodium.

Genitalia (Figs 12C,F-H): Apicolateral process elongate, thick, tapering near apex, with apical seta. Gonocoxite moderately elongate. Gonostylus thick basally, chelate-shaped, tapering to sharply pointed apex. Aedeagus with well developed basal apodemes, with tapering posterior extension. Paramere well developed, separate, tapering to sinuous, elongate apex, in one specimen apically bifurcate (Fig. 12H).

Distribution and bionomics. C. obuncus is known from amber collected 5 km south of Grassy Lake, Alberta (n= 2 ♂) and from the shore of Cedar Lake, Manitoba (n= 3 ♀) (Fig. 20B).

Because the female of *C. obuncus* is unknown, it is not possible to determine its host (or whether it fed at all).

Several specimens were associated with other organisms as follows: a spider (ROM CL-II-59; spider in separate piece in vial); an orthorrhaphous Brachycera and eight alate nymphs of *Mesozoicaphis canadensis* (two of these in separate piece in box) (CAS 1019); Ichneumonoidea and one male and one female Chironomidae (CAS 517). The association with chironomids confirms the likelihood of the proximity of aquatic habitats.

Taxonomic discussion. There is some question as to whether the males of *C. obuncus* are all conspecific. There are significant differences in the male genitalia which may or may not be an artifact of preservation and/or angle of observation. In specimens CAS 1019 and CAS 517, the apicolateral process appeared more slender (but still of constant diameter for most of its length) and the inner lobe of the gonostylus could not be seen.

Types. Holotype, male adult in amber piece on microscope slide, labelled "Holotype Culicoides obuncus Borkent, Cedar Lake, Manitoba, F.M.C. Coll., MCZ 9023" (MCZC); paratypes: 2 males from type locality, MCZ 9024 (MCZC), ROM CL-II-59 (ROME); 2 males from near Medicine Hat [= 5 km s. Grassy Lake], Alberta, CAS 517 (CNCI), CAS 1019 (CNCI).

Derivation of specific epithet. The name *obuncus* (bent in, hooked) refers to the distinctive shape of the gonostylus.

Culicoides annosus Borkent n. sp.

Diagnosis. Male: only Cretaceous species of Ceratopogonidae with a wing bearing only scattered macrotrichia on apical half of the wing, apicolateral processes which are more or less cylindrical and taper near apex, gonostylus with a thick, broad base tapering gradually to pointed apex. Female: unknown.

Description. Male adult. Descriptive statistics: see Tables 3-8.

Head: Eyes bare, narrowly separated dorsomedially. Single dorsomedial seta on vertex not visible. Antenna with well developed plume, basal foramen of pedicel not visible, 13 separate flagellomeres, flagellomeres 11-13 more elongate than preceding flagellomeres (Fig. 12J), first flagellomere with sensilla coeloconica (Fig. 12K). Mouthparts moderately elongate. Palpus with five segments, third segment moderately elongate, somewhat swollen at midlength, with capitate sensilla arranged in a shallow pit (Fig. 12L).

Thorax: Scutal apodemes not visible. Scutum pruinose, with scattered elongate setae, humeral pit not visible. Anapleural suture not visible. Katepisternum without setae.

Wing (Fig. 12I): Macrotrichia present on terminal wing veins, with some scattered on membrane at apex of wing, fine microtrichia present on all membrane. Macrotrichia on alula not visible. Cells r_1 and r_{2+3} not clearly visible but likely separate and short. M bifurcating distal to r-m.

Legs: Femora, tibiae slender. Legs lacking armature. Pair of thick setae on fore and midleg trochanter not visible. Midleg tibia without apical spur. Hindleg first tarsomere without thick basal spine, without palisade setae. Claws apically bifid. Without empodium.

Genitalia (Figs 12M-N): Apicolateral process elongate, thick, tapering near apex, with apical seta. Gonocoxite moderately short. Gonostylus thick basally, tapering to pointed apex. Aedeagus with well developed basal apodemes, with tapering posterior extension. Paramere apically slender, curved, bases not visible.

Distribution and bionomics. C. annosus is known from amber collected 5 km south of Grassy Lake, Alberta (n= 1 ♂) and from the shore of Cedar Lake, Manitoba (n= 1 ♂) (Fig. 20B).

Because the female of *C. annosus* is unknown, it is not possible to determine its host (or whether it fed at all).

Types. Holotype, male adult in amber on microscope slide, labelled "Holotype Culicoides annosus Borkent, Cedar Lake, Manitoba, Carpenter Coll., MCZ 6890" (MCZC); paratype: 1 male from near Medicine Hat [= 5 km s. Grassy Lake], Alberta, CAS 262 (CNCI).

Derivation of specific epithet. The name *annosus* (old, aged) refers to the Cretaceous age of this species.

Culicoides casei Grogan and Szadziewski

Culicoides? casei Grogan and Szadziewski 1988:809. Sayreville, Middlesex Co., New Jersey, USA. Upper Cretaceous.

Diagnosis. Male: only species of Cretaceous *Culicoides* with only a few macrotrichia on the wing membrane (situated near apex). Female: unknown.

Description. Male adult. Descriptive statistics: see Tables 9-14.

Head: Eyes bare, narrowly separated dorsomedially. Vertex with single dorsomedial seta present. Antenna with well developed plume, 13 separate flagellomeres, flagellomeres 11-13 more elongate than preceding flagellomeres, sensilla coeloconica not visible. Mouthparts short. Palpus with five segments, third segment somewhat ovoid, with numerous capitate sensilla in well defined pit.

Thorax: With anterior scutal apodemes not visible. Scutum bare, with scattered elongate setae, humeral pit not visible. Anapleural suture short, extending to posterior margin of anepisternal cleft. Katepisternum without setae.

Wing: Macrotrichia present along very apical margin of wing, fine microtrichia present on all membrane. Macrotrichia on alula not clearly visible but likely present. Cells r_1 and r_{2+3} well defined, elongate. M bifurcating distal to r-m.

Legs: Femora, tibiae slender. Legs lacking armature, except a pair of slightly thickened bristles at apex of most tarsomeres. Pair of thick setae on fore and midleg trochanter not visible. Midleg tibia without apical spur. Hindleg first tarsomere without thick basal spine, apparently (not clearly visible) without palisade setae. Claw apically bifid. Without empodium.

Genitalia (Fig. 13A): Apicolateral process elongate, slender, with apical seta. Gonocoxite short, thick. Gonostylus thick basally, tapering to pointed apex. Paramere well developed, separate, slender, bent subapically, with pointed apex. Aedeagus forming single median structure with short, tapering posterior extension, shape of very apex uncertain.

Distribution and bionomics. Culicoides casei is known from a single male in amber collected in Sayreville, Middlesex Co., New Jersey, USA. (Fig. 20A).

The presence of two associated Orthocladinae in the piece of amber containing *C. casei* (Grogan and Szadziewski, 1988) likely confirms the presence of local aquatic habitats (however, some larvae of extant Orthocladinae are found in soil).

Taxonomic discussion. The amber containing *C. casei* was further ground and polished (now 5X4X3 mm), allowing for some additional features to be observed or corrected. Grogan and Szadziewski (1988) reported that the thorax had humeral pits but I was unable to see these clearly and cannot confirm this observation. In addition, Grogan and Szadziewski (1988) reported the antennal ratio as 0.86, scutum with dense pubescence, and the claws as simple; my observations indicate otherwise (see description). My measurements of the antenna were taken as the total distance of the combined flagellomeres, which may explain the marked difference reported here for the antennal ratio.

Grogan and Szadziewski (1988) described the first tarsomere of the hindleg as having "a weak row of palisade setae". Although a row may possibly be present, the opaque leg was difficult to interpret. Scattered setae on the hindleg may superficially appear to be a row of palisade setae in some fossils because only those that are bordering the visible margin are apparent. Examination of the legs with reflected light suggested to me that the setae were actually scattered (as reported in the description above).

Grogan and Szadziewski (1988) also suggested that the tips of the gonostyli of the holotype were broken off. One gonostylus was actually visible at an oblique angle but the tip of the other was buried in membrane (or exudate), giving an appearance of having been broken distally (Fig. 13A).

Grogan and Szadziewski (1988) drew attention to the fact that the cerci were directed posteriorly from tergite 9. I consider this feature (also present in a number of Baltic amber Ceratopogonidae) to be an artifact of preservation (Borkent and Grogan, in press). In general, such posteriorly directed cerci are found in specimens which are in an excellent state of preservation (i.e. completely opaque and no evidence of deterioration) and this suggests that the original resin may have put pressure on the body cavity which forces the cerci into an unusual position.

The specimen was originally examined as a loose piece of amber. As sustained exposure to air decomposes amber, the specimen is now permanently mounted in Canada Balsam on a microscope slide.

Type. Holotype, male adult in amber on microscope slide, labelled "PU88892-A Culicoides casei, Type, No. PU88892, 27.iii.1966, Woodbridge Clay Member (Cenomanian), Raritan Formation, U. Cret., Clay pits E. of corner Main St. and Truman Ave., Sayreville, Middlesex Co., N.J., Gerard R. Case, amber containing 2 Chironomidae and 1 ceratopogonid, Princeton Department of Geology" (now at PMNH).

Derivation of specific epithet. The name *casei* was given by Grogan and Szadziewski (1988) in recognition of the collector of the holotype, G.R. Case.

Minyohelea Borkent New Genus

Type species: Minyohelea pumilis Borkent n.sp., the only included species.

Diagnosis: Only genus of Ceratopogonidae with the following combination of character states: three segmented palpus with only a single segment beyond the thick segment three; wing with R_1, Rs and r-m thick and darkly pigmented, radial veins distal to these absent (♂) or very faint (♀); antenna with 13 separate flagellomeres.

Description: Very small adults. Eye pubescence not visible, eyes narrowly separated dorsomedially. Vertex with single dorsomedial seta present. Male antennae with 13 separate flagellomeres, flagellomeres of nearly equal length except for slightly longer flagellomere 13 (Fig. 13E), sensilla coeloconica not present, plume somewhat sparse, basal foramen of pedicel not visible. Female antenna with 13 separate flagellomeres, flagellomeres of nearly equal length (Fig. 13H), sensilla coeloconica not visible. Proboscis very short, details of female mouthparts not visible. Palpus likely with 4 segments, only one present distal to segment 3 (bearing the capitate sensilla), pit not visible but if present it would be very small (Figs 13C,G). Anterior scutal apodemes not visible. Scutum with long setae, humeral pits not visible. Anapleural suture not visible. Katepisternum without setae. Legs (Figs 13J-L) slender, unarmed. Midleg trochanter with pair of thick setae. Midleg tibia without apical spur. Apex of hindleg tibia with 4-5 elongate setae, spur not visible but if present it is small. Hindleg first tarsomere straight, with palisade setae along entire length, without basal thick spine (Fig. 13L). Female claws equal, short, inner teeth not visible. Male claws apically bifid. Empodium not visible. Wing with fine microtrichia, without macrotrichia on membrane, venation reduced, without Cu, only very base of M present, $R_{1,\,Rs}$ and r-m thick, darkly pigmented, without R_{2+3}, $R4_{+5}$ in male (Fig. 13B), very faint $R4_{+5}$ and a possible single radial cell in female (Fig. 13F). Alula without macrotrichia. Male genitalia (Fig. 13D) short, without apicolateral process. Female genitalia with spermathecae not visible, sternite 9 not visible, cercus of moderate size.

Taxonomic Discussion: Only two other species of Cretaceous ceratopogonids other than *Leptoconops* and *Austroconops* have a reduced number of palpal segments. *Atriculi-coides incompletus* has a four segmented palpus but is clearly unrelated to *Minyohelea* (Figs 22D, 23D). The other species with a reduced number of palpal segments is *Leptohelea taimyrica* (Remm), described from Taimyr amber and known only as a single male. However, in that species (based on the published description), the basal flagellomeres are fused, the single radial cell is closed, M and Cu veins are evident and tergite 9 of the male genitalia has apicolateral processes. The features of *L. taimyrica* are discussed further below under "Phylogenetic Relationships of Cretaceous Ceratopogonidae".

Derivation of generic name: *Minyo* (small, short) *helea* (marsh dweller) refers to the small size of this Cretaceous ceratopogonid.

Minyohelea pumilis Borkent n. sp.

Diagnosis. Male and female: as for genus as this is the only included species.

Description. *Male adult*. As for genus as this is the only included species. Descriptive statistics: see Tables 3-8.

Genitalia (Fig. 13D): Apicolateral process absent. Gonocoxite moderately elongate. Gonostylus slender, slightly curved. Parameres, or possibly apex of aedeagus, forming two elongate rods. Aedeagus not otherwise visible.

Female adult. As for genus as this is the only included species. Descriptive statistics: see Tables 9-14.

Distribution and bionomics. *Minyohelea pumilis* is known from amber collected 5 km south of Grassy Lake, Alberta (n= 6 ♂, 1 ♀) (Fig. 20B).

Because the details of the female mouthparts of *M. pumilis* were not visible, it is not possible to determine its host (or whether it fed at all).

It is striking that six males of *M. pumilis* were found in one small piece of amber and were in a similar state of preservation. This suggests that the males of this species aggregated and may have formed swarms.

Taxonomic discussion. Males and females were associated through the shared presence of their small size, a 3 segmented palpus and reduced wing venation.

Although the male specimens are labelled Medicine Hat, Manitoba, the name of the province is clearly an error. The place of collection is in Alberta and this is confirmed by the name of the collector, P. Boston, who found amber in southern Alberta (McAlpine and Martin, 1969:827). That the type locality is truly in Alberta is confirmed by the presence of other specimens collected by P. Boston in 1963 near Medicine Hat [= 5 km S. Grassy Lake], Alberta (McAlpine and Martin, 1966:531.

Types. Holotype, male adult in amber on microscope slide, situated under largest coverslip, closest to the labels and further indicated by two arrows drawn on the coverslip, labelled "Holotype Minyohelea pumilis Borkent", "C.N.C. No. CAS 131 Ceratopogonidae (Diptera) nr Medicine Hat [actually = 5 km south of Grassy Lake], MAN [sic] 1963, P. Boston, Det. Phuoc T. Dang, CNC No. 21660" (CNCI); allotype, female adult in amber on microscope slide, labelled "Allotype Minyohelea pumilis Borkent", "nr. Medicine Hat [= 5 km s. Grassy Lake], Alta, May, 1973, McAlpine & Teskey, CAS 709 (CNCI); paratypes: 5 males labelled as for holotype, in same piece of amber as holotype (CNCI).

Derivation of specific epithet. The name *pumilis* (dwarfish) refers to the reduced number of palpal segments present in this species.

Palaeobrachypogon Borkent New Genus

Type species: Palaeobrachypogon vetus Borkent n.sp.

Diagnosis: Male: only genus of Ceratopogonidae in Cretaceous amber with no macrotrichia or with macrotrichia on the membrane of the very apical margin of the wing, with the first tarsomere of the hindleg with well developed palisade setae and a stout, well developed basal spine. Female: only genus of Ceratopogonidae in Cretaceous amber with elongate, equal claws on each leg.

Description: Moderate size adults. Eyes bare (see taxonomic discussion for possible exception), narrowly separated dorsomedially. Vertex with single dorsomedial seta present (Fig. 18L). Male antenna with 13 separate flagellomeres or with flagellomeres 2-4 fused, flagellomeres 11-13 longer than preceding flagellomeres (Fig. 16G), sensilla coeloconica on first flagellomere not visible, plume well developed, basal foramen of pedicel not visible. Female antenna with 13 separate flagellomeres, flagellomeres 9-13 longer than preceding flagellomeres (Figs 14I,15G,16I), first flagellomere with sensilla coeloconica (Fig. 15B). Proboscis moderately elongate (Figs 14B,15B). Labrum not visible. Mandible with at least 6 large teeth. Lacinia not visible. Palpus with five segments, segment 3 ovoid (Fig. 14B,15C) to moderately elongate (Fig. 15B), with capitate sensilla arranged on surface or in small, subapical pit. Hypopharynx not visible. Anterior scutal apodemes not visible. Scutum pruinose, with short to long scattered setae, humeral pits not visible. Anapleural suture short, extending to posterior margin of anepisternal cleft (Fig. 18K). Katepisternum without setae. Legs (Figs 14C-E,15D-F,16M-N) slender or with somewhat thicker hind femur and tibia or with all femora and tibiae somewhat thicker, unarmed except apex of each tarsomere of each leg with pair of thick setae. Midleg trochanter with pair of thick setae. Midleg tibia without apical spur. Apex of hindleg tibia with 6-7 elongate setae, well developed plumose spur. Hindleg first tarsomere straight, with palisade setae along entire length, with basal thick spine. Foreleg, midleg, hindleg claws (Figs 14G-H, 15I-M,16J-L) equal, moderately elongate, with very slender inner tooth arising from near base of each talon. Without empodium. Male claws apically bifid. Wing (Figs 14A,B,15A,16A,H) without or with few macrotrichia on membrane on apical margin of wing, very fine microtrichia on membrane, alula without macrotrichia, radial cells well developed, moderately elongate, cell r_{2+3} open in one species. M_2 faint or absent basally. M bifurcating distal to r-m. Male genitalia (Figs 14L-O,16B-F) of moderate size, with apicolateral process. Female genitalia with spermathecae not visible, sternite 9 not visible, cercus of moderate size (Fig. 14J).

Taxonomic Discussion: This new genus is distinct from all other genera of Ceratopogonidae through the combination of the following character states: first flagellomere with sensilla coeloconica, male antenna with separate flagellomeres 10 and 11, without katepisternal setae, wing without macrotrichia on membrane, well developed radial cells (second may be open), the first tarsomere of the hindleg with a complete row of palisade setae, females with long, equal claws on each leg, and legs more or less slender (some with somewhat thicker femora and tibiae).

Although the above combination of character states will allow for the recognition of the five included species, this in itself is weak evidence for the recognition of a new higher taxon; these should generally be proposed on the basis of an understanding of cladistic relationships. The current failure to do so, is a reflection of a combination of two features. First, our knowledge of the cladistic relationships between extant genera of Ceratopogonini is fragmentary. Second, in most instances it is uncertain whether extant genera are monophyletic units (see below under "Phylogenetic analysis"). Nevertheless, the fossil species included in *Palaeobrachypogon* cannot be readily placed in any extant genus. Until such time as further cladistic analysis is available for the extant fauna, the justification for recognizing the genus *Palaeobrachypogon* will be in question.

Two of the species described by Remm (1976) from Taimyr amber are here recognized as new combinations: *Palaeobrachypogon macronyx* (Remm), 1976:113 and *Palaeobrachypogon frigidus* (Remm), 1976:115. Both species have wings devoid of macrotrichia on the membrane and have elongate but equal female claws. They are similar in a number of features to the species described here from Canadian amber. It is possible that one or more of the species described here may actually be conspecific with one or both of the species from Taimyr amber. However, on the basis of the published descriptions, there appear to be significant differences between species of the two faunas and these are noted in the key below. Fortunately, R. Szadziewski (pers. comm.) is currently studying the Taimyr material and further observations may resolve some of the difficulties discussed here.

Palaeobrachypogon macronyx, P. frigidus and *P. aquilonius* were originally described in the genus *Ceratopogon* Meigen (Boesel, 1937; Remm, 1976). Without any supporting justification, Wirth and Grogan (1988) placed *P. macronyx* and *P. aquilonius* and Szadziewski (1988: 237) placed *P. frigidus* in *Brachypogon* Kieffer. Szadziewski (1988) retained both *P. macronyx* and *P. aquilonius* in *Ceratopogon*. Borkent and Grogan (in press) discussed details of the morphology of the latter two species which indicate that they cannot belong in that genus. Most importantly, both species have palisade setae along the entire length of the first tarsomere of the hindleg (see Fig. 14F: synapomorphy 28 in "Phylogenetic relationships of extant Ceratopogonidae").

Remm (1976) described *P. frigidus* as a species of *Fanthamia* Meillon (then as a subgenus of *Ceratopogon*) presumably because of the shared open second radial cell (R_1 is absent). This feature may be a valid synapomorphy of the genus, indicating the proper phylogenetic position of the fossil. However, the character state is also present in some other genera (i.e. *Meunierohelea*, some *Brachypogon*) suggesting either more inclusive relationship or the presence of homoplasy. Furthermore, *Fanthamia* also exhibits another synapomorphy not shared by the fossil: although some other Ceratopogonidae have reduced male antenna (plume and/or antennal segments) only male *Fanthamia* have a completely female-like antenna (including the different scape) and this suggests that it is a valid synapomorphy of the genus (see synapomorphy 38 under "Phylogenetic relationships of extant Ceratopogonidae").

Wirth and Grogan (1988) expressed doubt about *P. frigidus* actually belonging in *Fanthamia*. Szadziewski (1988: 90, 93), citing personal communications with Grogan and Wirth, placed *P. frigidus* in *Brachypogon* (*Brachypogon*) but suggested significant differences in the relative lengths of the female claws. However, the presence of separate male antennal flagellomeres in the fossil indicates that it cannot belong within *Brachypogon* or its closest relatives, all of which have the basal flagellomeres including flagellomere 11 fused (Grogan and Borkent,

1992). Until further comparisons and character state analyses have been made, there is, therefore, some question as to where *P. frigidus* belongs phylogenetically. In the interim, the species is placed in the new genus.

Remm (1976) suggested that both *P. macronyx* and *P. frigidus* have eyes with pubescence. Considering the discrepancies in the descriptions of other species (i.e. *Atriculicoides*), these too may be artifacts of observation.

The association between the sexes of the three species of *Palaeobrachypogon* recognized in Canadian amber is tenuous and needs to be tested when further material becomes available. The male and female of *P. remmi* were associated on the basis of their larger size. The male and female of *P. vetus* were associated on the basis of the more open radial cells, which in *P. aquilonius* are more slit-like. This left *P. aquilonius* without any recognized male.

Key to Fossil *Palaeobrachypogon*

The male of *P. aquilonius* is unknown. The distinctions between the species from Taimyr amber are based on the descriptions given by Remm (1976).

1. Wing with cell r_{2+3} open (R_1 absent). . . . .*frigidus* (Remm) (Taimyr amber, Upper Cretaceous)
1' Wing with cell r_{2+3} closed (R_1 present) 2
2. Male 3
2' Female 5
3. Gonostylus strongly bent at midlength (Figs 14L,O) . . . . *remmi* n.sp. (Canadian amber, Upper Cretaceous)
3' Gonostylus more or less straight or with slight curve (Fig. 16B,F) 4
4. Tergite 9 apically broad, with broadly spaced apicolateral processes (Fig. 16F). . . . . *vetus* n.sp. (Canadian amber, Upper Cretaceous)
4' Tergite 9 apically narrow, with narrowly spaced apicolateral processes . . . . . *macronyx* (Remm) (Taimyr amber, Upper Cretaceous)
5. Hind claws only slightly elongate with hind claw/hindleg fifth tarsomere length = 0.4-0.5; claws without inner teeth (Fig. 14G-H) . . . . *aquilonius* (Boesel) (Canadian amber, Upper Cretaceous)
5' Hind claws more elongate with hind claw/ hindleg fifth tarsomere length > 0.8 (Fig. 16L); claws with inner teeth (not known for *macronyx*) 6
6. Wing without macrotrichia on membrane (Fig. 16H); hindleg claws elongate (Fig. 15L), at least 1.5x the length of the fore and midleg claws . . . . *vetus* n.sp. (Canadian amber, Upper Cretaceous)
6' Wing with macrotrichia along apical margin of wing (Fig. 15A) (not known for *macronyx*); hindleg claws of moderate length (Figs.

15I,M), similar in size to fore and midleg claws ....
...........*remmi* n.sp. (Canadian amber, Upper Cretaceous)
.... *macronyx* (Remm) (Taimyr amber, Upper Cretaceous)

Derivation of generic name: Palaeo (old) *brachy* (short) *pogon* (beard) refers the structural similarity of the included species to extant members of *Brachypogon* and its closest relatives.

Palaeobrachypogon aquilonius (Boesel) NEW COMBINATION

Ceratopogon aquilonius Boesel, 1937:49. Cedar Lake, Manitoba. Upper Cretaceous.
Brachypogon aquilonius: Wirth and Grogan, 1988:31.

Diagnosis. Male: unknown. Female: only Cretaceous species of Ceratopogonidae without macrotrichia on the wing membrane, with claws of equal length on each leg and with the hind claw extending about 0.4-0.5 the length of the fifth tarsomere.

Description. Female adult. Descriptive statistics: see Tables 9-14.

Head: Eyes bare, dorsomedial separation not visible. Single dorsomedial seta on vertex not visible. Antennae with 13 separate flagellomeres, flagellomeres 11-13 more elongate than preceding flagellomeres (Fig. 14I), first flagellomere with sensilla coeloconica. Mouthparts moderately short (Fig. 14B). Membrane between clypeus and head capsule not visible. Labrum not clearly visible. Mandible with at least 6 large teeth. Lacinia not visible. Palpus with five segments, third segment ovoid, with capitate sensilla arranged in shallow depression or small pit (not clearly visible) (Fig. 14B). Hypopharynx not visible.

Thorax: Scutal apodemes not visible. Scutum bare, with scattered elongate setae, humeral pit not visible. Anapleural suture not visible. Setae on katepisternum not visible.

Wing (Fig. 14A): Macrotrichia present on membrane at very apical margin of wing, very fine microtrichia present on all membrane. Alula without macrotrichia. Cells r_1 and r_{2+3} well defined, moderately elongate. M bifurcating distal to r-m.

Legs (Figs 14C-E): Femora, tibiae moderately thick. Legs lacking armature, except for apex of each tarsomere of each leg with pair of thick setae. Fore and midleg trochanter each with pair of thick setae. Midleg tibia without apical spur. Hindleg first tarsomere with thick basal spine, with row of palisade setae along entire length (Fig. 14F). Foreleg, midleg, hindleg claws equal, moderately elongate, without inner teeth (Fig. 14G,H). Hindleg claw shorter than fifth tarsomere (Fig. 14H). Without empodium.

Genitalia (Fig. 14J): Spermathecae not visible. Sternite 9 not visible. Setae on segment 10 not visible. Cerci of moderate length.

Distribution and bionomics. P. aquilonius is known from amber collected 5 km south of Grassy Lake, Alberta (n= 1 ♀) and from the shore of Cedar Lake, Manitoba (n= 1 ♀) (Fig. 20B).

Female *P. aquilonius* likely fed on the blood of other insects (see below under "Evolution of blood-sucking and host selection in Ceratopogonidae"). The presence of somewhat enlarged claws suggests that they fed on prey which were of similar or smaller size (as do extant species with this feature).

Taxonomic discussion. The original placement of this species in *Ceratopogon* cannot be supported as discussed above under the taxonomic discussion of the genus and by Borkent and Grogan (in press). The subsequent placement of this species in *Brachypogon* by Wirth and Grogan (1988) may be a possibility because some critical features could not be observed (i.e. whether a seta is on the katepisternum; male features). However, the lack of katepisternal setae and fused male flagellomeres 2-11 in the morphologically similar *P. remmi* and *P. vetus* suggests that *P. aquilonius* does not belong in *Brachypogon*.

The amber containing the type specimen of *P. aquilonius* was further cut and polished to allow for more detailed observations. Therefore, although Boesel (1937) noted that the legs and abdomen were in a poor position, these could now be studied in more detail.

Although it was not labelled as such, it is clear the holotype was collected from Cedar Lake, Manitoba, where all early Canadian amber originated (including the other species which Boesel described).

Boesel (1937) described the relative lengths of the flagellomeres of *C. aquilonius* and noted that flagellomere 11 was not longer than some previous flagellomeres. I was able to observe that flagellomeres 11-13 were actually proportionally longer.

Although no male is recognized for *P. aquilonius*, it is possible that the male of either *P. vetus* or *P. remmi* is actually conspecific with *P. aquilonius*. The association of the sexes in these species is discussed above under the taxonomic discussion of the genus.

Type. Holotype, female adult in amber on microscope slide, labelled "Holotype Ceratopogon aquilonius No. 61, "Boesel"", "61", "Palaeobrachypogon aquilonius Det. A. Borkent" (ROME).

Material examined. 2 females: ROM 61 (ROME), 82-15-33 (TMPD).

Derivation of specific epithet. The name *aquilonius* (northern) given by Boesel presumably refers to the type locality in Canada.

Palaeobrachypogon remmi Borkent n. sp.

Diagnosis. Male: only Cretaceous species of Ceratopogonidae without macrotrichia on the wing membrane, with a strong basal spine and row of palisade setae on the first tarsomere of the hindleg, and a gonostylus strongly bent at midlength. Female: only Cretaceous species of Ceratopogonidae with macrotrichia restricted to apical margin of the wing membrane, with well developed moderately elongate radial cells, with a strong basal spine and row of palisade setae on the first tarsomere of the hindleg, with the hindleg claw elongate (hindleg claw/fifth tarsomere > 0.8) and with the fore, mid and hindleg claws more or less equal in size and with inner teeth.

Description. Male adult. Descriptive statistics: see Tables 3-8.

Head: Eyes bare, dorsomedial separation not visible. Single dorsomedial seta on vertex not visible. Details of antenna not visible. Mouthparts moderately short. Palpus with five segments, third segment moderately elongate, with capitate sensilla not visible.

Thorax: Scutal apodemes not visible. Scutum pruinosity not visible, with elongate setae, humeral pit not visible. Anapleural suture not visible. Setae on katepisternum not visible.

Wing (Fig. 14K): Lying at poor angle. Macrotrichia absent on membrane, very fine microtrichia present on all membrane. Alula without macrotrichia. Cells r_1 and r_{2+3} well define, short. Bifurcation of M not visible.

Legs: Femora, tibiae slender. Legs lacking armature. Pair of thick setae on fore and midleg trochanter not visible. Midleg tibia without apical spur. Hindleg first tarsomere with thick basal spine, with palisade setae along entire length. Claws apically simple. Without empodium.

Genitalia (Figs 14L-O): Apicolateral process elongate, moderately thick, cylindrical, with apical seta. Gonocoxite moderately elongate. Gonostylus thick basally, tapering to pointed apex, very bent at midlength. Only apex of paramere visible, slender, tapering to fine point. Only apex of aedeagus visible, apically bifurcate.

Female adult. Descriptive statistics: see Tables 9-14.

Head: Eyes bare, narrowly separated dorsomedially. Vertex with single dorsomedial seta present (Fig. 18L). Antennae with 13 separate flagellomeres, flagellomeres 9-13 more elongate than preceding flagellomeres (Fig. 15G), first flagellomere with sensilla coeloconica (Fig. 15B). Mouthparts moderately elongate, details not visible (Fig. 15B). Clypeus separated laterally by membrane from head capsule. Palpus with five segments, third segment ovoid to moderately elongate, with capitate sensilla not visible (Fig. 15C).

Thorax: With three anterior scutal apodemes. Scutum pruinose, with scattered elongate setae, humeral pit not visible. Anapleural suture short, extending to posterior margin of anepisternal cleft (Fig. 18K). Katepisternum without setae.

Wing (Fig. 15A): Macrotrichia present along apical margin of wing membrane, very fine microtrichia present on all membrane. Macrotrichia on alula not visible. Radial cell r_1 not visible, cell r_{2+3} well developed. M bifurcating distal to r-m.

Legs (Figs 15D-F): Femora, tibiae somewhat thick. Legs lacking armature, except single thick spine of first tarsomere of foreleg (visible only in MCZ 9029), apex of each tarsomere of each leg with pair of thick setae. Fore and midleg trochanter each with pair of thick setae. Midleg tibia without apical spur. Hindleg first tarsomere with thick basal spine, with palisade setae along entire length. Foreleg, midleg, hindleg claws (Figs 15I-M) equal, simple, with inner tooth arising from near base of each talon, claws from each leg of about equal length. Hindleg claw shorter than fifth tarsomere (Figs 15I,M). Without empodium.

Genitalia: Two spermathecae spherical in outline (Fig. 15H). Sternite 9 not visible. Setae on segment 10 not visible. Cerci of moderate length.

Distribution and bionomics. P. remmi is known from amber collected 5 km south of Grassy Lake, Alberta (n= 1 ♂) and from the shore of Cedar Lake, Manitoba (n= 3 ♀) (Fig. 20B).

Because the details of the female mouthparts of *P. remmi* were not visible, it is not possible to determine whether it fed or not. However, if it had functional mouthparts, the presence of somewhat enlarge claws would suggest that it sucked blood from other insects of similar or smaller size.

The holotype male of *P. remmi* is associated with a male *Atriculicoides globosus*. The amber including one female also held a female (?) chironomid (ROM 73) and another piece bearing a female *P. remmi* also included a twig of *Glyptostrobus* and a piece of what may be a liverwort (MCZ 9030). The single extant species of *Glyptostrobus* (Chinese swamp cypress) lives along the banks of streams and rivers in damp situations in southern China. Fossil *Glyptostrobus* are a common feature of Upper Cretaceous deposits in Alberta.

Taxonomic discussion. Males and females were associated through the shared presence of a wing at least mostly devoid of macrotrichia on the membrane, a long wing (relative to other *Palaeobrachypogon* in Canadian amber) and a first tarsomere of the hind leg with a stout basal spine and a row of palisade setae. The tenuous nature of this association is discussed further under the "Taxonomic Discussion" of the genus.

It may that more than one species is represented by the female material named as *P. remmi*. In particular, the relative sizes of the claws seemed to represent greater variation than one finds in extant species of Ceratopogonini. Nevertheless, it was unclear how to further distinguish the species without naming three species for three individuals. Clearly, further material would help to resolve this problem.

Types. Holotype, male adult in amber on microscope slide, labelled "Holotype Palaeobrachypogon remmi Borkent, nr. Medicine Hat [= 5 km s. Grassy Lake], Alta., June, 1974, Daze, Cooper, McAlpine, CAS 947b, CNC No. 21666" (CNCI); allotype, female adult in amber on microscope slide, labelled "Allotype Palaeobrachypogon remmi Borkent, Cedar Lake, Manitoba, F.M.C. Coll., MCZ 9029" (MCZC); paratypes: 2 females from Cedar Lake, Manitoba, MCZ 9030 (MCZC), ROM 73 (ROME).

Derivation of specific epithet. The name *remmi* is named in honor of the late Dr. H. Remm who contributed so much to our understanding of Palaearctic Ceratopogonidae and who provided an analysis of the Ceratopogonidae found in amber from the Taimyr Peninsula.

Palaeobrachypogon vetus Borkent n. sp.

Diagnosis. Male: only Cretaceous species of Ceratopogonidae with a very few macrotrichia on the apical margin of the wing membrane, with a strong basal spine and row of palisade setae on the first tarsomere of the hindleg, a more or less straight gonostylus and tergite 9 apically broad (apicolateral processes broadly separated). Female: only Cretaceous species of Ceratopogonidae without macrotrichia on the wing membrane, with well developed moderately elongate radial cells, with a strong basal spine and row of palisade setae on the first tarsomere of the hindleg, and with the hindleg claws shorter than the fifth tarsomere but at least 1.5x the length of the fore and midleg claws.

Description. Male adult. Descriptive statistics: see Tables 3-8.

Head: Eyes bare, narrowly separated dorsomedially. Vertex with single dorsomedial seta present. Antenna with well developed plume, basal foramen of pedicel not visible, 13 separate flagellomeres, flagellomeres 11-13 more elongate than preceding flagellomeres (Fig. 16G), first flagellomere with poorly visible sensilla coeloconica. Mouthparts moderately elongate. Palpus with five segments, third segment ovoid, with capitate sensilla arranged in small pit.

Thorax: Scutal apodemes not visible. Scutum pruinose, with scattered short and elongate setae, humeral pit not visible. Anapleural suture short, extending to posterior margin of anepisternal cleft. Katepisternum without setae.

Wing (Fig. 16A): Without macrotrichia or with a few macrotrichia on membrane at very apex of wing, very fine microtrichia present on wing membrane. Alula without macrotrichia. Cells r_1 and r_{2+3} well defined, moderately short. M bifurcation not clearly visible.

Legs: Femora somewhat thick, tibiae slender. Legs lacking armature, except for apex of each tarsomere of each leg with pair of thick setae. Fore and midleg trochanter each with pair of thick setae. Midleg tibia without apical spur. Hindleg first tarsomere with thick basal spine, with palisade setae along entire length. Claws apically bifid. Without empodium.

Genitalia (Fig. 16B-F): Apicolateral process elongate, tapering to pointed apex, with apical seta. Gonocoxite moderately short. Gonostylus thick basally, tapering gradually to pointed apex, more or less straight. Parameres separate, elongate, each tapering to pointed apex, apex curved ventrally. Aedeagus (Fig. 16E) of moderate size, basal apodemes not visible, with single posterior extension.

Female adult. Descriptive statistics: see Tables 9-14.

Head: Eye pruinosity not visible, dorsomedial separation not visible. Vertex with single dorsomedial seta present. Antennae with 13 separate flagellomeres, flagellomeres 9-13 more elongate than preceding flagellomeres (Fig. 16I), sensilla coeloconica on first flagellomere not visible. Mouthparts moderately elongate, details not visible. Membrane separating clypeus from head capsule not visible. Palpus not visible.

Thorax: Scutal apodemes not visible. Scutum pruinosity not visible, with scattered elongate setae, humeral pit not visible. Anapleural suture not visible. Setae on katepisternum not visible.

Wing (Fig. 16H): Without macrotrichia on wing membrane, microtrichia not visible. Macrotrichia on alula not visible. Cells r_1 and r_{2+3} well developed, moderately short. M bifurcating distal to r-m.

Legs (Figs 16M-N): Femora, tibiae slender or thick, with hind femur thicker than fore and midleg femora. Legs lacking armature, except for apex of each tarsomere of each leg with pair of thick setae. Pair of thick setae on fore and midleg trochanter not visible. Midleg tibia without apical spur. Hindleg first tarsomere with thick basal spine, with palisade setae apparently along entire length (not visible distally on one specimen; not visible at all in other). Foreleg, midleg, hindleg claws (Fig. 16J-L) equal, elongate, with inner tooth arising from near base of each talon, claws from hindleg longer than those of fore and midleg. Hindleg claw nearly equal in length to fifth tarsomere. Without empodium.

Genitalia: Spermathecae not visible. Sternite 9 not visible. Setae on segment 10 not visible. Cerci of moderate length.

Distribution and bionomics. P. vetus is known from amber collected 5 km south of Grassy Lake, Alberta (n= 1 ♂, 2 ♀) and from the shore of Cedar Lake, Manitoba (n= 2 ♂) (Fig. 20B).

Because the details of the female mouthparts of *P. vetus* were not visible, it is not possible to determine whether it fed or not. However, if it had functional mouthparts, the presence of somewhat enlarge claws would suggest that it sucked blood from other insects of similar or smaller size.

The holotype male was associated with an alate homopteran nymph and an unidentified mite.

Taxonomic discussion. Males and females were associated through the shared presence of a wing without macrotrichia on the membrane, a shorter wing, broader radial cells, and a strong basal spine and row of palisade setae on the first tarsomere of the hindleg.

The allotype female was in an awkward position for accurate observation and was partially enclosed in a sheath of air. The paratype female was in very poor condition, with only the legs in a reasonable state of preservation.

Types. Holotype, male adult in amber on microscope slide, labelled "Holotype Palaeobrachypogon vetus Borkent, nr. Medicine Hat [= 5 km s. Grassy Lake], Alta, June, 1974, McAlpine, Cooper & Daze, CAS 1198, CNC No. 21665" (CNCI); allotype, female adult in amber on microscope slide, labelled "nr. Medicine Hat [= 5 km s. Grassy Lake], Alta., May, 1973, McAlpine & Teskey, CAS 692" (CNCI); paratypes: 2 males from Cedar Lake, Manitoba MCZ 9031 (MCZC), ROM CL-II-58 (ROME); 1 female from Grassy Lake [= 5 km s. Grassy Lake], Manitoba, CAS 764 (CNCI).

Derivation of specific epithet. The name *vetus* (old) refers to the Cretaceous age of this species.

Peronehelea Borkent New Genus

Type species: Peronehelea chrimikalydia Borkent n.sp., the only included species.

Diagnosis: Male: only genus of Ceratopogonidae in Cretaceous amber with a few macrotrichia on the membrane at the apex of the wing and with the first tarsomere of the hindleg with well developed palisade setae and a stout, well developed basal spine. Female: only genus of Ceratopogonidae in Cretaceous amber with very elongate, unequal hind claws.

Description: Moderate size adults. Eyes bare, narrowly separated dorsomedially. Vertex with single dorsomedial seta present (Fig. 17F). Male antennae with 13 separate flagellomeres, flagellomeres 11-13 longer than preceding flagellomeres (Fig. 17B), sensilla coeloconica on first flagellomere not visible, plume well developed, basal foramen of pedicel not visible. Female antenna with 13 separate flagellomeres, flagellomeres 9-13 longer than preceding flagellomeres (Fig. 17I), first flagellomere likely with sensilla coeloconica. Proboscis moderately short (Fig. 17F). Labrum without lateral teeth. Mandible with about 13 large teeth (Fig. 17K). Lacinia not clearly visible but likely without teeth. Palpus with five segments, segment 3 moderately elongate, with about 6-7 capitate sensilla arranged in small, subapical pit (Fig. 17J). Hypopharynx smooth, without teeth. Anterior scutal apodemes not visible. Scutum pruinose, with short to long scattered setae, humeral pits not visible. Anapleural suture not visible. Katepisternum without setae. Legs (Figs 18A-D) slender except for somewhat thicker hind femur and tibia, unarmed except for 3-5 ventral spines on first tarsomere of foreleg, 2-3 ventral spines on first tarsomere of midleg, apex of each tarsomere of each leg with pair of thick setae. Midleg trochanter with pair of thick setae. Midleg tibia without apical spur. Apex of hindleg tibia with 6-7 elongate setae, well developed plumose spur. Hindleg first tarsomere straight, with palisade setae along entire length, with basal thick spine (Figs 18C-D). Female claws (Figs 18E-I) on fore and midleg equal, of moderate length, with inner teeth. Female claws on hindleg single, very elongate, with sinuous or straight inner tooth. Male claws apically bifid. Without empodium. Wing with macrotrichia on membrane near apex, very fine microtrichia on membrane, alula without macrotrichia, radial cells well developed, moderately elongate, M_2 faint basally, M bifurcating distal to r-m (Figs 17A,G-H). Male genitalia of moderate size, with apicolateral process (Figs C-E). Female genitalia with spermathecae not visible, sternite 9 not visible medially but without vaginal apodeme, cercus of moderate size (Fig. 18J). Eggs very elongate, rounded anteriorly and posteriorly (Fig. 18J).

Taxonomic Discussion: Although this taxon is distinctive within the Cretaceous ceratopogonid fauna, it is more difficult to distinguish *Peronehelea* from some extant genera. The genus is very similar to *Monohelea* Kieffer and related genera (see Wirth and Grogan, 1988). However, the combination of elongate hind claws and equal fore and midleg claws in the female and sensilla coeloconica on the first flagellomere are a unique combination within the family. It should be noted that the sensilla coeloconica could be barely seen on the first flagellomere of two females and this character state may in fact not actually be present. If not, *Peronehelea* may not be distinguishable from a number of extant genera with equal fore and midleg claws and unequal and very elongate hindclaws. This poor resolution is partially a reflection of our incomplete understanding of the cladistic relationships of extant genera traditionally placed in the Ceratopogonini.

Derivation of generic name: Perone (pin) *helea* (marsh dweller) refers to the long hind claws of females of this genus, a feature unique within the Cretaceous ceratopogonid fauna.

Peronehelea chrimikalydia Borkent n. sp.

Diagnosis. Male and female: as for genus as this is the only included species.

Description. Male adult. As for genus as this is the only included species. Descriptive statistics: see Tables 3-8.

Genitalia (Figs 17C-E): Apicolateral process very elongate, tapering gradually to pointed apex, with apical seta. Gonocoxite moderately elongate, slender. Gonostylus straight or curved, thick basally, tapering gradually to pointed apex. Aedeagus large, with well developed basal, well developed thick lateral apodemes, posteriorly directed apically forked medial extension. Paramere elongate, slender, apically rounded.

Female adult. As for genus as this is the only included species. Descriptive statistics: see Tables 9-14.

Eggs (Fig. 18J). Without any distinctive chorionic sculpturing.

Distribution and bionomics. Peronehelea chrimikalydia is known from amber collected 5 km south of Grassy Lake, Alberta (n= 3 ♂, 5 ♀) and from the shore of Cedar Lake, Manitoba (n= 1 ♂, 4 ♀) (Fig. 20B).

Female *P. chrimikalydia* likely fed on the blood of other insects (see below under "Evolution of blood-sucking and host selection in Ceratopogonidae"). The presence of an elongate hindclaw indicates that their hosts were of equal or smaller size and were caught in flight, as do extant ceratopogonids with this feature.

One female (MCZ 9028) had a small associated arthropod which was enclosed in a sheen of air, making further identification difficult.

Two males and one female were associated in one piece of amber (CAS 339) indicating that the species was at least locally common and that it may have aggregated in certain areas.

Taxonomic discussion. Males and females were associated through the shared presence of several character states: a wing with some macrotrichia on the apical portion of the wing, stout spines on the first tarsomere of the foreleg, and the first tarsomere of the hindleg with a large basal spine and a complete row of palisade setae. In addition, two males and one female were present in the same piece of amber (CAS 339).

There is some variation within this species that may indicate that more than one species is actually present. One male (in piece CAS 339) appeared to have a different form of aedeagus (Fig. 17D). However, the specimen was somewhat shrivelled and could not be clearly seen; the differences may be due to distortion and the difficulty of making accurate observations. There was also some differences in the relative length of the female hind claw and the shape of its inner tooth (Figs 18G-I). Similarly, there appeared to be substantial variation in the amount of wing membrane bearing macrotrichia (Figs 17G-H). Although this variation appeared to be continuous and was therefore interpreted as indicative of the presence of a single species, further material will likely indicate the presence of more than one species under this name.

The description of eggs within the abdomen of one female *P. chrimikalydia* (including one that is partially extruded) is the first description of ceratopogonid eggs in amber. Pierce (1959, 1966) previously recorded silicified fossil ceratopogonid eggs in Miocene nodules from California, U.S.A.

Types. Holotype, male adult in amber on microscope slide, labelled "Holotype Peronehelea chrimikalydia Borkent, nr. Medicine Hat [= 5 km s. Grassy Lake], Alta., May, 1973, McAlpine & Teskey, CAS 882, CNC No. 21664" (CNCI); allotype, female adult in amber on microscope slide, labelled "nr. Medicine Hat [= 5 km s. Grassy Lake], Alta., July 8-11, 1971, J.F. McAlpine, CAS 339b" (CNCI); paratypes: 2 males, labelled as for allotype, CAS 339a (CNCI); 1 male, Cedar Lake, Manitoba MCZ 9025 (MCZC); 1 female, near Medicine Hat [= 5 km s. Grassy Lake], Alberta, 1963, P. Boston, CAS 228 (CNCI); 1 female, Grassy Lake [= 5 km s. Grassy Lake], Alberta, 1979, 79-16-8 (TMPD); 2 females, Grassy Lake [= 5 km s. Grassy Lake], Alberta, 82-15-74 (TMPD), 82-15-114 (TMPD); 3 females, Cedar Lake, Manitoba MCZ 9026 (MCZC), MCZ 9027 (MCZC), MCZ 9028 (MCZC); 1 female, Cedar Lake, Manitoba, August, 1963, J.F. McAlpine & J.E.H. Martin, CAS 523 (CNCI).

Derivation of specific epithet. The name *chrimikalydia* refers to the author's three children, Christopher, Michael and Lydia who show a great deal of interest and a reasonable level of tolerance of this entomologist's endeavors!

QUESTIONABLE FOSSIL CERATOPOGONIDAE

The Upper Cretaceous *Simulidium priscum* has been considered as a possible member of *Leptoconops* (Grogan and Szadziewski, 1988). However, the fossil, long thought to be lost, has been rediscovered in the BMNH and is actually a member of the Brachycera (Ross, pers. comm.). Evenhuis (1994:117) placed the fossil in the fossil family Protopleciidae.

Craig (1977), citing personal communications with W. Hennig, J.F. McAlpine, and R.W. Crosskey, indicated that *Pseudosimulium humidum* (Westwood, in Brodie, 1845) most likely was a ceratopogonid. This Lower Cretaceous (Norris, 1985) compression fossil has an overall habitus somewhat reminiscent of a ceratopogonid but, as noted by Grogan and Szadziewski (1988), there are no published details which would confirm its placement within the family. However, neither can another placement be suggested at the present time.

Zhang (1986) described *Luanpingites flavus* Zhang as a new species and genus in a new family Luanpingitidae, based on a single female. He indicated that this Middle Jurassic compression fossil from China was intermediate between Simuliidae and Ceratopogonidae, but that "it bears a closer affinity to Simuliidae". The basis for this placement was unclear. Somewhat suspect evidence suggests that this fossil species may be most closely related to the Thaumaleidae.

Thaumaleidae are the only members of the Culicomorpha in which the females have holoptic eyes and this is likely a synapomorphy of the group. Although some Forcipomyiinae and members of *Dasyhelea* have broadly contiguous eyes, these are secondarily derived and differ in detail. In addition to the above synapomorphy, Thaumaleidae have short antennae (about as long as the head) with only the distal 5-6 flagellomeres elongate. *Luanpingites flavus* shares both these features and this evidence indicates that the species may be the sister group of the Thaumaleidae. The presence of an apparently well developed proboscis is a plesiomorphy which excludes the fossil from the Thaumaleidae and the strikingly short tarsomeres are likely a synapomorphy of *L. flavus*.

Crosskey (1987) noted that *Simulia pasithea* Heyden (1870:238) and *Simulia terribilis* Förster (1891:467) may also be ceratopogonids but the descriptions of these specimens are far too inadequate to confirm these suggestions. Both these species are from German, Oligocene deposits and need to be re-examined to determine their true identity.

Ashe (1983) provided a synopsis of generic names applied to fossil Chironomidae. He noted that a number of these are so poorly known that they require re-examination before their familial identifications can be possibly determined or confirmed. As such, at least some of these may actually belong in the Ceratopogonidae; only study of the original material could possibly give a conclusive answer.

Finally, Hong (1981:52) described *Aspinus orientalus* Hong as a new genus and species of Chironomidae in Eocene amber from China (now in the genus *Amberaspinus* Evenhuis). The four elongate terminal male flagellomeres and the overall habitus indicate that the species is actually a ceratopogonid (and likely a *Forcipomyia*). Re-examination of the specimen is required to reappraise especially the wing venation (which is reported to have no radial cells and a simple vein M. The genitalia appears to have a simple anal point (modified tergite 9) which would be characteristically chironomid, but this may possibly be a misinterpreted aedeagus.

PHYLOGENETIC ANALYSIS

Phylogenetic Relationships of Extant Ceratopogonidae

To properly interpret the phylogenetic relationships of fossil members of a given taxon, it is vital that the cladistic relationships of its extant members be understood (Wiley, 1981). In the following section, some basal lineages of Ceratopogonidae are interpreted to allow for a further interpretation of Cretaceous Ceratopogonidae and to provide a basis for comparison with fossil ceratopogonid faunas described from the Tertiary.

To provide a cladistic analysis, I compared character states within the family to those of all other Culicomorpha families. I paid closest attention to the Chironomidae as the sister group of Ceratopogonidae (Borkent and McKeever, 1990; Sinclair, 1992a; Wood and Borkent, 1989). Within the Ceratopogonidae, I attempted to examine nearly every genus on a worldwide basis and studied many species in many of these (especially of genera in the Nearctic Region).

Previous analyses: Borkent et al. (1987) presented cladistic evidence interpreting the relationships between the subfamilies of Ceratopogonidae and further investigation confirms that interpretation (given below). Further analysis of the subfamily Ceratopogoninae was provided by Borkent and Grogan (in press) who suggested that a group of three genera, *Culicoides, Paradasyhelea* and *Washingtonhelea* Wirth and Grogan formed the sister group of the remaining Ceratopogoninae and of those remaining genera, that *Ceratopogon* was the sister group of the balance. These arguments form the basis for further analysis here and are summarized, discussed and somewhat modified below.

Previous phylogenetic interpretation by Remm (1975) was discussed in detail by Borkent et al. (1987).

Szadziewski (1988) provided an apparent cladistic interpretation of the Ceratopogonidae using a combination of synapomorphies and 'underlying synapomorphies' which was modelled on a method employed by Saether in his publications (e.g. Saether, 1970, 1977; Saether and Halvorsen, 1981). This approach provides, at best, a weak basis for interpreting phylogenetic relationships. Saether (loc. cit.), using cladistic terms in his analyses of Chironomidae and Chaoboridae (Borkent, 1993), provides detailed evolutionary trees based on his unjustified interpretation of character state polarities. Although described as plesiomorphic and apomorphic, character states are merely listed, without any explanation of the basis for polarity decisions. Explicit character state distributions are not discussed and without such information, clearly one of the most central and basic tenets of cladistic analysis (Watrous and Wheeler, 1981; Wiley, 1981), it is impossible to appraise Saether's basis for character state interpretation. The reader is left without any evidence for judging the legitimacy of the polarity decisions. Borkent (1979:196-201; 1984:139) and Cranston and Humphries (1988) have discussed details of character state interpretation provided by Saether in his analyses.

In following Saether's approach, Szadziewski (1988) also generally failed to provide character state distributions for most of the features he used to interpret the Ceratopogonidae. However, at least some of the polarity decisions are discussed explicitly. All of these are discussed below and some provide good evidence of cladistic relationship.

Szadziewski (1988) also apparently used Saether's (1983) concept of 'underlying synapomorphies' in constructing an evolutionary tree of the Ceratopogonidae. This concept (Saether, 1979a, 1979b, 1983, 1986, 1988) is based on the use of parallelisms in several taxa as evidence of a shared genetic inclination toward the evolution of a given character state and hence as evidence of close genealogical relationship (i.e. synapomorphy). Saether's (1983) application of underlying synapomorphies in cladistic analysis has been soundly criticized both in theory and in practice (Cranston and Humphries, 1988; Farris, 1986). In essence, although it is clear that parallel evolution occurs in many related groups of organisms, to use such patterns for the interpretation of cladistic relationship is unsupportable. Furthermore, the character state distributions of apparent 'underlying synapomorphies' are not discussed in any detail; again, by not making the data available, the conclusion of synapomorphy cannot be examined.

The following provides an analysis of character states important in interpreting phylogenetic relationships between the earlier lineages of Ceratopogonidae (not including many genera of Ceratopogonini which could not be examined firsthand and those of the monophyletic grouping composed of the Heteromyiini + Sphaeromiini + Palpomyiini + Stenoxenini). The synapomorphies are included in the cladogram in Fig. 21.

1. Genital fork (sternite 9) of female genitalia well developed, with vaginal apodeme extending along dorsal wall of common oviduct (plesiomorphic); vaginal apodeme very reduced or absent (apomorphic).

This character was discussed by Borkent et al. (1987) (as synapomorphy 1).

2. Larval pharyngeal apparatus present and acting as a sieve or absent; not acting as a crushing structure (plesiomorphic); pharyngeal apparatus a grinding and sifting structure (apomorphic).

This character was discussed by Borkent et al. (1987) (as synapomorphy 2).

3. Male adult tergite nine without apicolateral process or, if present, lacking setae (plesiomorphic); pair of apicolateral processes present and each bearing at least one seta (apomorphic).

Most families of Nematocera lack any specialized posterior projections on tergite nine. Articulated lobes are present in Ptychopteridae (Wood, 1991), setose lobes are present in many Culicidae (Harbach and Knight, 1980; Wood, 1991) and some Chaoboridae have a pair of bare projections (Cook, 1956). Within the Chironomoidea, a few Thaumaleidae have a pair of lobes arising from tergite nine (Theischinger, 1986; pers. obs.) and some Chironomidae, which appear to be derived lineages within the family, have a single, median projection (= anal point of chironomid specialists) (Saether, 1980). Of these two families, only some of the chironomids with a median projection have these bearing setae (usually a number). The Simuliidae have no projections arising from tergite nine.

Therefore, the presence of apicolateral processes may be homologous to those present in only some members of the following families: Chironomidae, Thaumaleidae, Chaoboridae, Culicidae and Ptychopteridae but I consider this unlikely. It seems more plausible that these are independently derived.

The presence of at least one seta on the apicolateral process in ceratopogonids appears to be a valid synapomorphy. Setae on projections of tergite nine are present in only some derived lineages of Chironomidae and in some Culicidae and these are likely due to homoplasy.

The apicolateral processes have been lost in a number of lineages within the Ceratopogonidae but are present in at least some members of all early lineages, with the exception of the monotypic *Austroconops* Wirth and Lee where they appear to be absent.

4. Larva with three thoracic and nine (or 10) abdominal segments (plesiomorphic); larva with body segments secondarily divided (apomorphic).

This character was discussed by Borkent et al. (1987) (as synapomorphy 5).

5. Female wing with radius joining costa without thickening (plesiomorphic); female wing with radius joining costa as a strong thickened stigma (Fig. 1F) (apomorphic).

The derived condition is unique within the Culicomorpha.

6. Female cerci short, rounded in lateral view (plesiomorphic); cerci elongate, laterally compressed (Figs 1L-M) (apomorphic).

The condition is unique in the Ceratopogonidae, and within the Culicomorpha otherwise occurs only in members of the Telmatogetoninae

(Saether, 1977), which are clearly specialized (marine) members of the Chironomidae. Szadziewski (1988:246) considered the derived state to be a synapomorphy of the entire genus. The elongate cerci of some *Leptoconops* are likely used to oviposit in sand (see further discussion under the genus above).

7. Third palpal segment with capitate sensilla in a pit (plesiomorphic); third palpal segment with capitate sensilla arising directly from surface of segment (apomorphic).

The interpretation of this character is weak. Both conditions occur in the sister group of *Leptoconops* (i.e. all other Ceratopogonidae, see Fig. 21) and varies intragenerically in a number of genera of Ceratopogoninae. The sister group of the Ceratopogonidae is the Chironomidae (Wood and Borkent, 1989; Borkent and McKeever, 1990) and the earliest lineage within this family is represented by the only member of the Chironomidae with biting mouthparts, the genus *Archaeochlus* Brundin. The females of this genus have a deep sensory pit on the third palpal segment (Cranston, et al., 1987). Otherwise within the Chironomidae, both character states are present. All Simuliidae have well developed palpal pits and Thaumaleidae lack the pits entirely.

In spite of this apparent homoplasy, I have interpreted the presence of capitate sensilla on the surface of the palpal segment as derived. Considering the high likelihood that synapomorphy 6 is correctly interpreted, ingroup comparisons (i.e. the presence of a well developed pit in species of the subgenera *Styloconops* and *Brachyconops*) suggest that the absence of the palpal pit in *Megaconops, L.* (*Leptoconops*) and *Proleptoconops* is derived.

8. Gonostylus with stout apical spine (plesiomorphic); gonostylus without apical spine (apomorphic).

As argued by Szadziewski (1988:246), the plesiomorphic feature is present in early lineages of Chironomidae and all Simuliidae (including *Parasimulium* Malloch, if the spine on its inner gonostylus is considered as homologous; Wood and Borkent, 1982). The spine is absent in Thaumaleidae and is either present or absent in the Culicoidea. These outgroup comparisons suggest that the absence of a stout seta on the apex of the gonostylus is derived within the Ceratopogonidae but may be susceptible to homoplasy.

9. Setae on vertex of adult head capsule scattered or in dorsolateral arrangement (plesiomorphic); in addition to other setae on vertex, a single seta located medially, just dorsal to where the eyes meet or, in groups where the eyes are separated dorsomedially, between these (apomorphic).

This character was discussed by Borkent et al. (1987) (as synapomorphy 3).

10. Postoccipital ridge of adult head capsule poorly sclerotized and tentorium directed in an anteroventral-posterodorsal angle (plesiomorphic); postoccipital ridge well developed and tentorium at a horizontal plane (apomorphic).

This character was discussed by Borkent et al. (1987) (as synapomorphy 4).

11. Male adult with apically simple claws (plesiomorphic); male with apically bifid claws (apomorphic).

Males of other families of Culicomorpha all have apically simple claws indicating that this feature is likely derived within the Ceratopogonidae.

Within the group defined by the synapomorphy, there are a few exceptions: *Echinohelea* Macfie and a few members of the Heteromyiini + Sphaeromiini + Palpomyiini + Stenoxenini (e.g. *Heteromyia* Say, *Clinohelea* Kieffer). Considering their position in the cladogram based on other synapomorphies (Fig. 21), these instances of apically simple claws must be considered to be secondary losses.

12. Adult midleg tibia with spur (plesiomorphic); midleg tibia lacking spur (apomorphic).

Within the Nematocera many groups have a pair of spurs at the apex of the midleg tibia (as well as on the fore and midleg). Within the Culicomorpha midleg spurs are present or absent in the Culicoidea. They are absent in Thaumaleidae and a pair is present in Simuliidae and in most Chironomidae (present in all early lineages).

Within the Ceratopogonidae, only members of *Leptoconops* possess a single midtibial spur. The spur is absent in all other Ceratopogonidae. Although reported as absent in *L. floridensis* Wirth (Wirth, 1951), a small spur is actually present (Wirth, pers. comm.).

This pattern indicates that the feature is likely derived within the Ceratopogonidae.

13. Adult eyes broadly separated dorsomedially (plesiomorphic); eyes moderately or narrowly separated (apomorphic).

Within the Culicomorpha, the Culicoidea have moderately to broadly separated eyes dorsomedially. Both male and female Thaumaleidae and most male Simuliidae (but not the earliest lineage, *Parasimulium*) have broadly contiguous eyes. Female Simuliidae have broadly separated eyes and the Chironomidae, as the sister group of the Ceratopogonidae (Wood and Borkent, 1989), have moderately to broadly separated eyes.

Within the Ceratopogonidae, only *Leptoconops* have broadly separated eyes. In other Ceratopogonidae, the eyes are more closely approximated. In those taxa with eyes which are moderately separated (e.g. *Paradasyhelea, Baeohelea, Fanthamia*, many members of the Heteromyiini + Sphaeromiini + Palpomyiini + Stenoxenini), the configuration of the frons and vertex appears different from that in *Leptoconops*. In many Chironomidae and *Leptoconops*, the vertex and frons are indistinguishable, with continuous medial cuticle present between the eyes (in many chironomids, there is an additional pair of elevated bumps bearing proprioceptive sensilla; Willassen, 1983). In Ceratopogonidae other than *Leptoconops*, the vertex (which bears the single medial seta) and the frons (lying between the antennal scapes) are almost always separate. The few exceptions are a few species of *Culicoides* in which the females have continuous cuticle between the vertex and frons; however, the males of these species have a distinct separation.

These observations suggest that the close or moderately close eyes present in all Ceratopogonidae other than *Leptoconops* is a derived condition. The corresponding separation of vertex and frons in this monophyletic group suggests that the moderately separated eyes of some Ceratopogonidae is likely a secondary feature derived from the more closely approximated eyes present in early lineages within the family (i.e. *Austroconops, Washingtonhelea, Culicoides* (see synapomorphy 18 regarding the condition in *Forcipomyia, Atrichopogon* Kieffer and *Dasyhelea*).

14. Adult thorax with posterior margin of scutellum rounded in dorsal view (plesiomorphic); posterior margin forming nearly a 90° angle (apomorphic).

This character was discussed by Borkent et al. (1987) (as synapomorphy 8).

15. Adult thoracic anapleural suture well developed, extending to anterior margin of anepisternal cleft (plesiomorphic); anapleural suture short, extending to posterior margin of anepisternal cleft (apomorphic).

This character was discussed by Borkent et al. (1987) (as synapomorphy 6).

16. Adult trochanter of fore and midleg with only thin, simple setae (plesiomorphic); trochanter of fore and midleg each with pair of thick, contiguous setae (apomorphic).

This character was discussed by Borkent et al. (1987) (as synapomorphy 7).

17. Male adult antennal pedicel with large basal foramen (plesiomorphic); pedicel with moderately-sized basal foramen (apomorphic', = 17'); pedicel with narrow basal foramen (apomorphic", = 17").

This feature was discussed by Borkent et al. (1987:596) as a character which was difficult to interpret. Comparison with other Chironomoidea showed that they have the most derived condition (17") but that both conditions were present in the Culicoidea, indicating some homoplasy.

Considering the distribution of other character states within the basal lineages of Ceratopogonidae (i.e. the strong evidence that *Leptoconops* and *Austroconops* are the earliest extant lineages within the family), it is likely that the large basal foramen is plesiomorphic. As such the moderately-sized basal foramen is derived (17') and the narrow basal foramen a further modification (17").

Examination of the monotypic genus *Washingtonhelea* indicated that it has the intermediate condition.

18. Adult eyes broadly separated to barely contiguous dorsomedially (Fig. 2H) (plesiomorphic); eyes broadly contiguous dorsomedially (Figs 4H,5D) (apomorphic).

The distribution of the plesiomorphic character state was discussed under synapomorphy 13.

In all species of *Forcipomyia* and *Atrichopogon* the eyes broadly abut medially with a dividing, longitudinal line of cuticle (or line of weakness) about 3-4 eyes diameters in length. In many (most?) *Dasyhelea*, the eyes also abut broadly, including those species in the *D. cincta* group, which likely represents the earliest lineage within the genus (Borkent, 1991). In those species of *Dasyhelea* where the eyes are separated by a small distance, the medial line of weakness is still present. In other Ceratopogonidae with closely or moderately separated eyes a line of weakness is not evident. This likely signifies the previous presence of broadly contiguous eyes in those *Dasyhelea* which now have somewhat separated eyes.

The broadly contiguous eyes are nearly unique within the Ceratopogonidae and considering outgroup comparisons (see synapomorphy 13), is likely to be derived. Especially for the females, the only other Culicomorpha with broadly contiguous eyes are the Thaumaleidae (which have many more ommatidia abutting medially). In addition, some *Parabezzia* Malloch and some *Stilobezzia* Kieffer also have broadly contiguous eyes medially. These similarities are likely due to homoplasy.

19. Female adult claws with inner tooth (plesiomorphic); claws simple, lacking inner tooth (apomorphic).

Both the plesiomorphic and apomorphic conditions are found in other Culicomorpha. Sliver-like projections are present in Thaumaleidae and only ornithophilic species of Simuliidae have a large basal tooth. Although both present and absent in the Chironomidae, the early lineages of this family have claws with inner teeth (Brundin, 1966:77; Cranston et al., 1987).

Within the Ceratopogonidae, virtually all early lineages have an inner tooth but in many this is reduced to a small sliver-like projection (e.g. *Leptoconops*, *Austroconops* (Borkent, et al., 1987), *Washingtonhelea*, *Culicoides*). *Forcipomyia*, *Atrichopogon* and *Dasyhelea* lack an inner tooth and this is interpreted to be a derived condition.

A few other Ceratopogonidae have also lost the inner tooth, indicating that the feature is susceptible to homoplasy: some *Ceratopogon* (Borkent and Grogan, in press), some *Macrurohelea* Ingram and Macfie, all *Parabezzia* and some *Palpomyia* Meigen.

20. Larval thoracic and abdominal setae arising from more or less flat areas of cuticle (plesiomorphic); setae arising from cuticular projections (apomorphic).

This character was discussed by Borkent et al. (1987) (as synapomorphy 9).

21. Terminal adult antennal flagellomere tapering gradually to apex (plesiomorphic); terminal antennal flagellomere with apical nipple which is basally constricted (apomorphic).

The derived condition is present in all *Forcipomyia* and *Atrichopogon* and is unique within the Culicomorpha.

22. Adult thoracic paratergite without setae (plesiomorphic); paratergite with 1-6 setae (apomorphic).

No other families of Culicomorpha have setae on the paratergite. Within the Ceratopogonidae only *Atrichopogon, Phaenobezzia* Haeselbarth and *Stenoxenus* Coquillett have setae on the paratergite. However, the condition in the latter two genera is almost certainly not homologous to that in *Atrichopogon* for two reasons. First, both *Phaenobezzia* and *Stenoxenus* Coquilett are members of the Palpomyiini and Stenoxenini respectively and are only distantly related to *Atrichopogon* (Fig. 21). Second, in *Phaenobezzia* and *Stenoxenus* the thorax is densely covered with setae including the lateral sclerites. The paratergite has setae on it as part of a broader modification (i.e. entire thorax covered with setae). In *Atrichopogon* the lateral thoracic sclerites are devoid of setae.

23. Male adult flagellomeres lacking longitudinal striations (plesiomorphic); flagellomeres with longitudinal striations (apomorphic).

This character was discussed by Borkent et al. (1987) (as synapomorphy 10).

24. Male and female adult scape lacking ventral apodeme (plesiomorphic); scape with ventral apodeme (apomorphic).

This character was discussed by Borkent et al. (1987) (as synapomorphy 11).

25. Egg ovoid to elongate (plesiomorphic); egg c-shaped (apomorphic).

The condition in *Dasyhelea* is unique within at least the Culicomorpha. Although eggs have been reported for only a few species of *Dasyhelea* (Keilin, 1921; Williams, 1944), examination of mature eggs in the abdomens of adult females of many species cleared on microscope slides indicates that the feature is widespread within the genus (pers. obs.).

26. Sternite 9 of female terminalia forming a continuous band ventrally (plesiomorphic); sternite 9 discontinuous medially, forming two halves (apomorphic).

Since the publication of the evidence for the interpretation of these character states (Borkent et al., 1987, as synapomorphy 12), R. Szadziewski (pers. comm.) pointed out that there are some exceptions to the derived state in a few species of *Culicoides* and some *Alluaudomyia* Kieffer.

The suggested monophyly of *Culicoides* and the phylogenetic position of *Alluaudomyia* indicates that a complete sternite 9 in only some species in each of these genera is best interpreted as secondarily reverted to the plesiomorphic state.

Alternatively, if *Culicoides* is not monophyletic it is possible, considering the phylogenetic position of the genus in relation to other genera, that the complete sternite 9 of a few species reflects a phylogenetic position originating before the synapomorphy (and hence that *Culicoides* is a paraphyletic group). However, the close morphological similarity of other features of these species (i.e. wing coloration, male genitalia) to other species of *Culicoides* also suggests that their complete sternite 9 is an example of homoplasy.

A recent re-examination of *Washingtonhelea* revealed that this monotypic genus also has a thin but complete sternite 9 which encompasses the opening of the spermathecal duct medially. The stippled structure shown by Wirth and Grogan (1988:149) in the approximate position of sternite 9 is actually the thickened posterior margin of sternite 8.

27. Female antennal flagellum without sensilla coeloconica or, if present, restricted to first flagellomere (plesiomorphic); flagellum with sensilla coeloconica on at least three flagellomeres (apomorphic).

The distribution of sensilla coeloconica on the first flagellomere of the female antenna among taxa was discussed by Borkent et al., (1987:600) and Felippe-Bauer and Bauer (1990) (also see below). Their presence in at least some non-ceratopogonids (in Corethrellidae, some Culicidae and Chironomidae) suggests that they (or sensilla in modified form) are plesiomorphic to the family. Most ceratopogonids have either sensilla coeloconica restricted to the first flagellomere or lack them altogether.

Virtually all species of *Culicoides* have at least three flagellomeres with sensilla coeloconica, with some always present on the first flagellomere; some have them on all 13 and most have one of a variety of combinations of 4-12 flagellomeres with sensilla coeloconica. There are two exceptions to the derived condition. The Nearctic *C. floridensis* has no sensilla coeloconica (Blanton and Wirth, 1979) and the Palaearctic *C. cameroni* has been reported having sensilla coeloconica restricted to the first flagellomere (Campbell and Pelham-Clinton, 1960). This may indicate that either of these two species (or both) are the sister group to all remaining *Culicoides* (unlikely) or these have secondarily lost sensilla on more terminal flagellomeres.

Only a few other genera of Ceratopogonidae have more than one flagellomere with sensilla coeloconica. *Austrohelea* have 2-3, *Fanthamia* have four with sensilla coeloconica, one species of *Macrurohelea* and a few species of *Brachypogon* have 5. Based on the evidence of other synapomorphies (Fig. 21), I consider these species to have acquired more extensively distributed sensilla coeloconica independently.

Females of the closely related *Paradasyhelea* (Fig. 21) apparently have 1-4 flagellomeres with sensilla coeloconica (Clastrier, 1989; Wirth, 1981). The presence of a number of flagellomeres with sensilla coeloconica in at least some members of this genus may indicate that the derived condition is actually a synapomorphy of *Culicoides* + *Paradasyhelea*. Further study is required.

The character state is more complex in *Culicoides* than the above discussion would suggest. All species of *Culicoides* have sensilla coeloconica on the first flagellomere. Those with a total of three flagellomeres with sensilla coeloconica, have the remaining sensilla on flagellomeres 12 and 13 or on flagellomeres 11 and 12. Of those with only a total of four, some have the remaining three on flagellomeres 6-8, 9-11, or 11-13. This may suggest that at least in these instances, the character state may not be homologous. However, it is important to know the flagellomeres are all part of a single segment which is secondarily divided. Therefore, the presence of sensilla coeloconica on different flagellomeres may be viewed as the shifting of sensilla positions within a single segment.

Understanding the evolution and significance of the sensilla coeloconica is further complicated by the fact that they represent only one structural form of sensilla that also occur in other Ceratopogonidae; sensilla coeloconica are one shape of sensilla that may possibly otherwise be in a deep pit (called ampullacea), somewhat protruding (called basiconica) or more elongate (called trichodea). The following early lineages within the Ceratopogonidae have different sensilla in the same location on the first flagellomere as the sensilla coeloconica found in *Culicoides* and numerous other Ceratopogonini: *Leptoconops* (trichodea, ampullacea), *Austroconops* (trichodea), *Forcipomyia* and *Atrichopogon* (trichodea, basiconica), and *Dasyhelea* (trichodea). It is highly conceivable that these sensilla types have been modified from one type to another and that a given form is therefore a questionable indicator of phylogenetic relationship. Wirth and Navai (1978) have previously indicated as well, that sensilla trichodea, basiconica and coeloconica grade into one another and that sensilla ampullacae are merely sunken sensilla coeloconica.

Finally, it has recently become apparent that the recognition of such sensilla as either trichodea, basiconica or coeloconica does not accurately portray their function and that the presence of no pores, one pore or many pores in the cuticular surface is a better indicator of role (and therefore, possibly, as homologous structures) (Blackwell, et al., 1992b; Frazier, 1985). Unfortunately, we presently lack comparative information about the state of such pores for most ceratopogonids; details have only been described for a few species of *Culicoides* (Blackwell, et al., 1992b; Chu-Wang, et al., 1975).

In light of the above information, I consider the described synapomorphy only as weak evidence of the monophyly of *Culicoides* and it would be highly desirable to discover further evidence of the monophyly of this economically important group.

28. First tarsomere of adult hindleg lacking palisade setae (plesiomorphic); first tarsomere with partial row of palisade setae (apomorphic', = 28'); first tarsomere with complete row of palisade setae (apomorphic", = 28").

This character was discussed by Borkent and Grogan (in press).

29. Male adult aedeagus with medial prong absent, a single lobe or bifurcate at very apex (plesiomorphic); aedeagus with medial portion divided into two elongate prongs (apomorphic).

This character was discussed by Borkent and Grogan (in press, as synapomorphy 1).

30. Adult hind tibial spur present (plesiomorphic); hind tibial spur reduced (apomorphic).

This character was discussed by Borkent and Grogan (in press, as synapomorphy 2).

31. Male adult parameres separate or fused but not overlapping (plesiomorphic); parameres expanded and overlapping near midlength (apomorphic).

This character was discussed by Borkent and Grogan (in press, as synapomorphy 3). There is some homoplasy within *Ceratopogon*.

32. Adult katepisternal setae absent (plesiomorphic); katepisternum with at least one seta (apomorphic).

This character was discussed by Grogan and Borkent (1992) (as synapomorphy 1).

33. Male adult antennal flagellomeres separate (plesiomorphic); flagellomeres 5-11 fused (apomorphic', = 33'); flagellomeres 2-11 (apomorphic", = 33").

This character was discussed by Grogan and Borkent (1992) (as synapomorphy 2).

34. Female adult labium, fleshy, with setae arising directly from cuticle (plesiomorphic); labium truncated with most apical setae on cuticular extensions (plesiomorphic).

The condition present in the monotypic *Rhynchohelea* Wirth and Blanton is unique within the Culicomorpha and hence considered derived. It is important to note, however, that the male of *Rhynchohelea* is unknown and therefore could not be scored for synapomorphy 28.

35. Adult anepisternal setae absent (plesiomorphic); anepisternum with 2-3 setae along posterior margin (apomorphic).

This character was discussed by Borkent (1992:436).

36. Male genitalia with parameres (plesiomorphic); genitalia lacking parameres (apomorphic).

All basal lineages of Ceratopogonidae and members of other families of Culicomorpha have parameres (Wood, 1991; Wood and Borkent, 1982). Within the Ceratopogoninae, two other genera are also lack parameres (Wirth and Grogan, 1988) and I consider these similarities to be due to homoplasy: *Camptopterohelea* Wirth and Hubert and *Fittkauhelea* Wirth and Blanton (neither genus seen firsthand) (Szadziewski and Grogan, 1994a).

37. Adult palpus with five segments (plesiomorphic); palpus with two segments (apomorphic).

The apomorphic condition is unique within Ceratopogonidae. All other families of Culicomorpha have five palpal segments with only some groups of Chironomidae (derived lineages) and Culicidae having a reduced number (Szadziewski and Grogan, 1994a).

A few other Ceratopogonidae have only 3-4 palpal segments (listed in Borkent et al., 1987: 598; also see synapomorphy 48 below) but, based on other relationships (Fig. 21), these are not likely to represent intermediate states.

Szadziewski and Grogan (1994a) also supported this character state as evidence of the sister group relationship between *Baeohelea* and *Baeodasymyia* Clastrier and Raccurt.

38. Male antenna plumose, scape lacking dorsal flange (plesiomorphic); antenna with short hairs, scape with dorsal flange abutting against ventral margin of frons (apomorphic).

The apomorphic state in *Fanthamia* is unique within the Ceratopogonidae and represents the acquisition of a female-like antenna. The presence of a dorsal flange on the scape is characteristic of females of many Ceratopogonidae (including *Fanthamia*) but is not present in males of other genera.

A few other Ceratopogonidae have a reduced male antennal plume (Borkent and Grogan, in press) but these still retain at least some other male characteristics (relative length of terminal flagellomeres, some setae still elongate, scape without dorsal flange, enlarged pedicel).

It is possible that the reduced antennal plume in male *Parastilobezzia* Wirth and Blanton (not seen) is also an instance of an acquired female antenna.

39. Aedeagus with continuous, pigmented cuticle medially (plesiomorphic); aedeagus membraneous medially with two pigmented, lateral halves (apomorphic).

The apomorphic condition appears to be unique within the Culicomorpha. However, *Isthmohelea* Ingram and Macfie may also have a split aedeagus (Wirth and Grogan, 1988) but no specimens were seen firsthand; if so, this would be considered as homoplasy as the genus likely is closely related to other genera with medially continuous aedeagae (Wirth and Grogan, 1988).

At least one *Stilobezzia* has a small separate medial sclerite at the apex of aedeagus, in addition to the lateral halves (i.e. *S. bulla* Thomsen). This may indicate that this species is either the earliest lineage within *Stilobezzia* (and hence suggesting that the presence of medial membrane is actually a derived condition within the genus) or a reacquisition of a condition somewhat similar to the plesiomorphic character state. The latter interpretation seems more probable as *S. bulla* is likely closely related to *S. thomsenae* Wirth, a species lacking any medial pigmentation.

40. Female adult abdomen directed posteriorly, with short ninth and tenth segments (about as long as cerci) (plesiomorphic); abdomen bent forward ventrally, with ninth and tenth segments markedly elongate (apomorphic).

The derived condition is unique within the Culicomorpha.

41. Female wing tapering gradually to rounded apex, setae of more or less uniform size and shape along apex of wing (plesiomorphic); wing indented apically with large, stout setae along indentation (apomorphic).

The derived condition is unique within the Culicomorpha.

42. Male adult cerci small, located ventrally on membraneous patch which is itself ventral to tergite 9 (plesiomorphic); male cerci large, laterally compressed, directed posteriorly beyond margin of tergite 9 (apomorphic).

The derived condition is unique within the Culicomorpha.

Although there is some variation in the size of male cerci in earlier lineages, all of these originate on the ventral surface of tergite 9. In the group with the derived condition, the cerci are strikingly enlarged, are laterally compressed and originate on the posterior margin of tergite 9, somewhat resembling female cerci in size and shape.

The only possible exceptions within the group defined by this synapomorphy are members of *Echinohelea*, which have strikingly modified genitalia (Wirth and Grogan, 1988). Tergite 9 lies between the markedly enlarged gonocoxites as a narrow strip. The cerci are strongly sclerotized, somewhat circular in cross-section but do project partially past the apex of tergite 9 as lobes. The cerci are large in comparison to the width of tergite 9 and each extends anterolaterally. I consider the modifications found in *Echinohelea* to be further derived from the apomorphic condition described above.

43. Adult hindleg femur more or less slender and bare of stout bristles or, femur may be stout and if bristles present, these arranged in scattered pattern or in one row (plesiomorphic); hind femur swollen with 2-4 rows of strong bristles (apomorphic).

This character was discussed by Borkent and Bissett (1990:208).

44. If 2-4 rows of strong bristles present on femur of male hindleg, then these restricted to less than 0.58 of the femur length (plesiomorphic); bristles extending more than 0.58 the length of the male hind femur (apomorphic).

This character was discussed by Borkent and Bissett (1990:209).

45. Female adult with clypeus separated by membrane from margin of eye (plesiomorphic); clypeus fused with ventromedial margin of eye (apomorphic).

This character appears to be unique within the Culicomorpha and is a complex and convincing indicator of the relationship between the genera possessing it.

46. Female adult wing with costa ending at vein R4+5 (plesiomorphic); costa extending beyond apex of R4+5 (apomorphic).

The apomorphic character state is nearly unique within the Ceratopogonidae; it is also present in a few species of *Atrichopogon* (Wirth and Ratanaworabhan, 1993). Female *Parastilobezzia* likely also have the derived condition and although they are placed here in the undefined group represented by "Other Genera" in Fig. 21, they may therefore actually belong within the clade. Unfortunately, I was unable to examine specimens of *Parastilobezzia* firsthand and could not determine the condition of a number of other character states pertinent to its phylogenetic placement (i.e. characters 42, 45).

Otherwise within the Culicomorpha, the derived state occurs in all Thaumaleidae, some early lineages of Simuliidae and to a slight degree, in a few Chironomidae.

47. Adult wing with two radial cells (plesiomorphic); wing with one radial cell (apomorphic).

The interpretation of this character is poorly supported. Within the earliest lineages of Ceratopogonidae the two cell condition is prevalent: the wing of *Leptoconops* is too modified to allow for the character to be scored and *Austroconops* has two well developed cells. The Forcipomyiinae have two cells (very reduced in some species) and *Dasyhelea* Kieffer (Dasyheleinae) appear to have only one, although the cells are so slit-like that they too are difficult to interpret. A few species of *Culicoides* also have a one cell condition. Finally, all conditions are present within *Paradasyhelea*: completely closed, one cell and two celled species.

Within the Ceratopogonini, the one cell condition is otherwise found in the following genera: *Afrohelea* Wirth, *Alluaudomyia, Baeodasymyia, Baeohelea*, some *Brachypogon*, some *Camptopterohelea, Ceratoculicoides* Wirth and Ratanaworabhan, *Fanthamia, Leptohelea* Wirth and Blanton, *Nannohelea* Grogan and Wirth, *Paralluaudomyia* Clastrier, *Rhynchohelea, Schizonyxhelea* Clastrier, some *Stilobezzia*.

Within the Heteromyiini + Sphaeromiini + Palpomyiini + Stenoxenini, the one cell condition is also present in a number of genera: some *Amerohelea* Grogan and Wirth, *Bezzia* Kieffer, *Calyptopogon* Kieffer, *Ceratobezzia* Kieffer, *Chelohelea* Giles and Wirth, *Crispomyia* Debenham, *Macropeza* Meigen, *Neobezzia* Wirth and Ratanaworabhan, some *Nilobezzia* Kieffer, *Niphanohelea* Grogan and Wirth, *Paryphoconus* Enderlein, *Pellucidomyia* Macfie, *Phaenobezzia, Probezzia* Kieffer, *Stenoxenus, Tetrabezzia* Kieffer, *Xenohelea* Kieffer.

The above character state distribution shows that the proposed apomorphic state is widely susceptible to homoplasy. Furthermore, it is clear from this and further studies (Borkent, in prep.) that many of the genera listed above are most closely related to taxa with the two cell condition.

Nevertheless, examination of the earliest lineages of Ceratopogonidae indicate that the two cell condition is likely to be plesiomorphic within the family. The presentation of the one cell state, however, can only be regarded as very weak evidence for considering *Neurohelea* Kieffer and/or *Heteroceratopogon* Wirth and Grogan as the sister group of *Bahiahelea* Wirth + *Fittkauhelea* + *Diaphanobezzia* Ingram and Macfie + *Parabezzia*.

48. Adult palpus with five segments (plesiomorphic); adult palpus with four segments, with segments four and five fused (apomorphic).

The presence of a five segmented palpus is widespread within the Culicomorpha with only a few Culicidae and Chironomidae (clearly young lineages) having a reduced number.

Within the Ceratopogonidae, virtually all genera have the five palpal segments. There are only a few exceptions.

A few genera of very tiny ceratopogonids exhibit a palpus which is correspondingly reduced (*Baeohelea* and *Baeodasymyia* (2 segmented; discussed under synapomorphy 37); *Leptohelea* and *Nannohelea* (3 segmented, with only one past the sensilla bearing segment); the Upper Cretaceous *Minyohelea*, described herein).

In *Camptopterohelea* and some species of *Culicoides* the palpal segments are reduced in number but there are still two segments distal to the sensilla bearing third segment (i.e. the basal segments cannot be identified), indicating that the reduction cannot be considered homologous to the apomorphic state described here.

As discussed by Borkent et al. (1987), the earliest two lineages of

Ceratopogonidae (*Leptoconops, Austroconops*) exhibit a fusion of the fourth and fifth palpal segments. However, considering the series of nested sets of apomorphies which separate these from the taxa under consideration here, these must have been independently derived.

A few lineages within *Forcipomyia* (e.g. Debenham and Wirth 1984; Debenham, 1987a, 1987b), one species of the fossil genus *Atriculicoides* (*A. incompletus*) and a few species of *Culicoides* (e.g. *C. pygmaeus* Tokunaga) also have the derived feature. Other members of both these genera have the plesiomorphic condition, suggesting that these too, are due to homoplasy.

Finally, species of *Stenoxenus* have palpal segments four and five fused but, considering their sister group relationship with *Paryphoconus* Enderlein, a genus with separate segments, and the phylogenetic position of the Stenoxenini (Fig. 21), this feature has been acquired independently.

This character state distribution suggests that the apomorphic condition is correctly interpreted, although somewhat susceptible to homoplasy.

49. Male adult parameres moderately developed, separate or, if fused, forming flat medial structure (plesiomorphic); parameres very large and fused into large sphere (apomorphic').

The apomorphic state is unique in the Culicomorpha.

50. Male adult with segment 9 variously developed but not longer than genitalia (plesiomorphic); segment 9 very large (apomorphic).

The large segment 9 of *Bahiahelea* is virtually unique in the Ceratopogonidae. Only a few species of *Atrichopogon* also have a very large segment 9, which must certainly be due to homoplasy.

51. Male fore coxa with normally developed setae similar to those found on mid and hind coxa (plesiomorphic); fore coxa with markedly developed, stout setae (apomorphic).

The apomorphic state is unique in the Ceratopogonidae.

52. Anterior margin of male adult scutum with three ventrally directed apodemes arising from separate bases and widely separated from one another (Fig. 19A-C) (plesiomorphic); anterior margin with three apodemes closely associated medially and arising from common base (Fig. 19D) (apomorphic', = 52'); anterior margin with two apodemes arising from broadly developed base (Fig. 19F), in some with very short medial third apodeme (Fig. 19E) (apomorphic", = 52").

The plesiomorphic condition is present in all members of the Ceratopogonidae other than those defined by the synapomorphy. In *Leptoconops, Austroconops,* Forcipomyiinae and *Dasyhelea* the apodemes are thin and sometimes short. Members of the Ceratopogonini have the apodemes more strongly developed as do both *Neurohelea* and *Neurobezzia* Wirth and Ratanaworabhan (I recently collected male

Neurobezzia in association with females; their identification and characterization by Grogan and Wirth (1978) is incorrect).

The intermediate apomorphic condition is present only in adult *Echinohelea* (Fig. 19D. The most derived condition is present in all members of the Heteromyiini + Sphaeromiini + Palpomyiini + Stenoxenini. This character would be characteristic for both sexes except that female *Clinohelea* have three fully developed apodemes, somewhat approximating the plesiomorphic condition. However, the presence of the apodemes on a common base and the presence of the derived condition in the male, indicates that the character state in the female is likely a reversal.

The condition in other Culicomorpha is difficult to interpret. The apodemes are absent in Culicoidea, Thaumaleidae and Simuliidae. At least some Chironomidae (*Archaeochlus* Brundin, *Buchonomyia* Fittkau, a variety of Podonominae) have the lateral apodemes situated at the very anterolateral margin of the scutum but appear to lack the medial apodeme.

This pattern may suggest that the three apodeme condition is a synapomorphy of the entire family (with subsequent modifications).

53. Wing with M_1 and M_2 forking beyond crossvein r-m, with petiolate base (plesiomorphic); M_1 and M_2 forking at or just basal to r-m, lacking petiolate base (apomorphic', = 53'); M_1 and M_2 forking well before r-m, lacking petiolate base (apomorphic", = 53").

Virtually all early lineages of Ceratopogonidae have wings in which M_1 and M_2 fork beyond crossvein r-m. Those that do not are lacking the base of or the entire M_2 (i.e. *Rhynchohelea*) making it impossible to score the character state at all.

In the presentation here, the intermediate apomorphic state has evolved at least three times; in *Neurohelea, Neurobezzia* and in the lineage *Echinohelea* + Heteromyiini + Sphaeromiini + Palpomyiini + Stenoxenini. In addition, the base of M_2 is very near crossvein r-m in *Heteroceratopogon*. Considering the close relationship between *Neurohelea, Neurobezzia* and *Heteroceratopogon* (Fig. 21), it is possible that the intermediate condition evolved at the same node as synapomorphy 45 and that the lineage defined by synapomorphy 48 + *Bahiahelea* reverted to the plesiomorphic condition.

The most derived condition defines members of the Heteromyiini + Sphaeromiini + Palpomyiini + Stenoxenini. However, even within this assemblage many (but not all) species of *Bezzia* possess the intermediate condition. Female adult *Bezzia* share a complex derived condition of modified tergal abdominal apodemes (Borkent and Craig, 1994) with other Palpomyiini, which provides good evidence that this tribe is monophyletic. Other members of the tribe have the fully derived condition of lacking the petiolate base of vein M_2. This evidence suggests that some members of *Bezzia* have reverted to the intermediate apomorphic condition (or that the fully derived condition has evolved at least twice within the Palpomyiini).

54. Gonocoxites separate (plesiomorphic); gonocoxites fused basally both dorsally and ventrally (apomorphic).

The derived condition is unique in the Culicomorpha.

The above synapomorphies form a nested set as illustrated in the cladogram presented in Fig. 21[1].

This cladogram presents several implications for our understanding of derived character states within the family, the phylogenetic relationships between genera of Ceratopogonidae and their classification.

The cladogram in Fig. 21 shows that the Forcipomyiinae (including *Forcipomyia* and *Atrichopogon*) and Dasyheleinae (including *Dasyhelea*) form a monophyletic group. These two taxa were also presented as sister taxa by Szadziewski (1988:246, his Trend 4) on the basis of four character states (two of these without any further discussion or justification; one discussed further above as synapomorphy 18). The fourth was supported with an analysis with which I disagree. Szadziewski (1988) indicated that the presence of an anterior larval proleg in Forcipomyiinae and Dasyheleinae was a derived feature because it is absent in the larvae of *Leptoconops* (the earliest lineage of Ceratopogonidae; see Fig. 21). He considered the presence of a proleg in first instar *Culicoides* as independently derived from that in Forcipomyiinae + Dasyheleinae.

Outgroup comparisons indicate that the proleg of these ceratopogonid taxa is actually plesiomorphic. All other Chironomoidea have an anterior proleg, as do the first instar larvae of Corethrellidae (Borkent and McKeever, 1990). Furthermore, the apex of the proleg bears numerous curved hooks which look similar in all these taxa (Barr, 1984; Borkent and McKeever, 1990; Nielsen, 1951:49). This evidence strongly suggests that the derived feature is plesiomorphic within the Chironomoidea and possibly also within the Culicomorpha (Borkent and McKeever, 1990).

The absence of a proleg in *Leptoconops* is likely correlated with its very modified larval morphology, adapted to living in wet sand.

The loss of the larval proleg may be a synapomorphy grouping all Ceratopogonidae other than Forcipomyiinae, *Dasyhelea*, and *Culicoides* (with an independent loss in *Leptoconops*). However, the first instar larvae are unfortunately unknown for the vast majority of Ceratopogoninae and until further information is available, the character state cannot be further interpreted. This hypothesis would indicate that the larvae of

[1] Szadziewski and Grogan (1994) have recently provided good evidence for the monophyly of both *Baeohelea* and *Baeodasymyia*. Although they suggested that the absence of the fore tibial spur in *Baeodasymyia* was unique in the family, it is also absent in *Dasyhelea*. Nevertheless, considering strong evidence that *Dasyhelea* belongs elsewhere phylogenetically, this similarity must be due to homoplasy.

Austroconops and *Washingtonhelea*, once they become known, will also have an anterior larval proleg.

When Wirth (1952) first proposed formal names for some of the subfamilies and tribes of Ceratopogonidae, the division between the Ceratopogonini and Stilobezziini seemed clearly defined. Subsequent description of the world fauna, particularly of genera from the southern hemisphere, resulted in increasing difficulty in applying the concept of these two tribes. Wirth and Grogan (1988) recognized that there was no character or group of characters which could be used to justify their status. In addition they emphasized that there was no clear understanding of the cladistic relationships between members traditionally included in the Ceratopogonini and the Stilobezziini. They therefore combined the two tribes under the name Ceratopogonini. Although no synapomorphies were proposed in support of this revised concept, the presence of a single tribe at least allowed for the recognition of its members.

As a consequence of this action the subfamily Ceratopogoninae is now divided into the following six tribes: Culicoidini, Ceratopogonini, Heteromyiini, Sphaeromiini, Palpomyiini, and Stenoxenini. The last four tribes have long been thought to be closely related and have been named the *Palpomyia-Bezzia* group (Johannsen 1943; Macfie, 1940), the Stenoxenini (Wirth, 1952) and the Palpomyiinae (Remm, 1975; Tokunaga, 1966). In the present study the group is referred to simply as Heteromyiini + Sphaeromiini + Palpomyiini + Stenoxenini.

The tribe Culicoidini has been considered to include three genera: *Culicoides*, *Paradasyhelea*, and *Austroconops* (Szadziewski, 1988; Wirth et al., 1974). Borkent et al. (1987) showed that *Austroconops* actually belonged elsewhere as an early lineage of the family (Fig. 21). More recently, Borkent and Grogan (in press) indicated that the genus *Washingtonhelea* was phylogenetically close to *Culicoides* and *Paradasyhelea* (as an unresolved polychotomy including the ancestor of all other Ceratopogoninae). The present analysis shows that *Washingtonhelea* actually forms the sister group to all remaining Ceratopogoninae. *Culicoides* and *Paradasyhelea* remain as unresolved taxa.

The three synapomorphies proposed by Szadziewski for the Culicoidini (1988:249; his Trend 9) (and in which he included *Austroconops*) cannot be presently supported. The first, male "parameres usually with long, slender and curling tip", is not present in *Paradasyhelea*, is absent in some species of *Culicoides*, and furthermore, is present in a number of other Ceratopogonidae including at least the following: some *Forcipomyia*, *Washingtonhelea*, some *Ceratoculicoides*, some *Alluaudomyia*, and some *Stilobezzia*. This character state requires further investigation, as it is possible that the derived state is a synapomorphy of the genus *Culicoides*, with some subsequent reversals within the genus and some homoplasy in other genera.

The second character state suggested by Szadziewski (1988) to support the monophyly of the Culicoidini was the male "tergite IX usually long and bearing distinct, slender and pointed apicolateral processes". I see no difference between most *Culicoides* and many other Ceratopogonidae in this regard (e.g. some *Dasyhelea*, some *Forcipomyia, Paradasyhelea*, some *Ceratoculicoides*, some *Ceratopogon*, and some *Alluaudomyia*).

The third character was the presence of "3 pairs of chromosomes". Szadziewski (1988) noted that this character was poorly known and his conclusion was based on the following information: n= 3 in *Culicoides* and n= 4 in *Dasyhelea, Stilobezzia, Sphaeromias* Curtis and *Bezzia*. However, Atchley and Jackson (1968) reported n= 3 for *Nilobezzia* and *Probezzia*. *Palpomyia* have n= 4 (Issayev, 1983). The picture is further clouded by outgroup comparisons. The Chironomidae have n= 3 or 4, Simuliidae have n= 2 or 3, Thaumaleidae have n= 3, Corethrellidae have n= 3, Culicidae have n= 3 or 4 and Chaoboridae have 4. Clearly, this feature is homoplastic and likely a poor indicator of relationship.

As such, these character states provide no evidence that either *Culicoides* or *Paradasyhelea* are monophyletic or that they form a monophyletic group together; the concept of the Culicoidini remains suspect. A poorly supported synapomorphy for *Culicoides* is proposed above as synapomorphy 27.

Although most papers devoted to the study of Ceratopogonidae concern themselves with analysis of some aspect of species of *Culicoides* and though many species in the genus are of immense economic, medical and veterinary importance, there are no cladistic analyses of the genus yet available. This is most unfortunate, as such an evolutionary framework would provide a logical basis for comparison of the wealth of bionomic, ecological and distributional data available on at least some members of this genus. Furthermore, such an analysis would possibly provide a framework within which to interpret the fossils described in this study.

Downes (1978:58, see also pg. 29) provided a list of character states for the Heteromyiini + Sphaeromiini + Palpomyiini + Stenoxenini but their exact distribution amongst taxa was not provided. The cladistic analysis here (and Borkent, in prep.) indicate that many of those character states do not clearly define the Heteromyiini + Sphaeromiini + Palpomyiini + Stenoxenini but actually define more, or less, inclusive monophyletic groups (e.g. see character 45 above).

The conclusions in this study point to a pattern which additional data (Borkent, in prep.) confirms. The Ceratopogonini is a grade concept, with the youngest lineages being sister taxa to the Heteromyiini + Sphaeromiini + Palpomyiini + Stenoxenini. Although the Ceratopogonini could be defined and recognized (e.g. Wirth and Grogan, 1988), the character states used to do so were a combination of synapomorphies shared with the Heteromyiini + Sphaeromiini + Palpomyiini + Stenoxenini (to separate them from older lineages of Ceratopogonidae) and plesiomorphies in relation to the Heteromyiini + Sphaeromiini + Palpomyiini + Stenoxenini (to separate them from this younger group). The cladistic analysis here suggests that the past difficulties in defining the tribe Ceratopogonini were real because the group does not actually exist as a monophyletic taxon.

The placement of *Neurohelea* and *Neurobezzia* in the cladogram requires comment. These genera were formally placed in the Heteromyiini and their transfer has implications for our understanding of the Heteromyiini. When Grogan and Wirth (1979b) proposed a phylogeny of the genera of Heteromyiini they presented two character states which argued for the inclusion of *Neurohelea* in the tribe. The first was the synapomorphy defining the tribe, a swollen fifth tarsomere on the foreleg. This character state does indeed appear to be unique within the Ceratopogonidae and is therefore logically interpreted as a synapomorphy. However, re-examination of specimens shows that the character state is not or weakly present in *Neurohelea* or *Neurobezzia* and that the fifth tarsomeres of these two genera do not differ significantly from a number of genera in the Ceratopogonini. Furthermore, the lack of a swollen fifth tarsomere in *Tetrabezzia*, a genus clearly closely related to *Pellucidomyia* and *Heteromyia* on the basis of the modified hind leg, indicates that the character is susceptible to reversal.

The second character state presented as a synapomorphy of *Neurohelea*, *Neurobezzia* and *Physohelea* Grogan and Wirth is the same as the intermediate synapomorphy 53' described above. Synapomorphy 53" was presented as grouping all remaining Heteromyiini. There were several difficulties in this presentation. Firstly, as an intermediate state, the presence of M2 arising from near crossvein r-m cannot be used to group the basal lineages of Heteromyiini. The character state logically groups all of the Heteromyiini. Secondly, both the intermediate and completely apomorphic character states are present in other taxa outside the Heteromyiini, as indicated in the discussion of synapomorphy 53 above.

I therefore consider that there is no evidence arguing against the current placement of *Neurohelea* and *Neurobezzia*.

The only synapomorphy for the Heteromyiini (without *Neurohelea* and *Neurobezzia*) is the enlarged fifth tarsomere of the foreleg.

Szadziewski (1988) suggested that his analysis of Ceratopogonidae showed that the classification employed by Wirth should be inverted, with the Ceratopogoninae being "the most plesiomorphic" and the Forcipomyiinae + Dasyheleinae as "most apomorphic". However, as argued by Borkent and Grogan (in press), his concluding phylogeny supports the contention that *Leptoconops,* Forcipomyiinae, *Dasyhelea* and *Culicoides* are all early lineages within the family. The degree to which taxa are modified (i.e. autapomorphies) should be inconsequential to the determination of their classification.

The discovery that the Ceratopogonini are paraphyletic indicates a need for a reappraisal of the formal classification of the family. However, I refrain from making any proposals until further studies are complete of further characters of adults and immatures (Borkent, in prep.) and suggest that the current higher classification be retained for the time being. Further studies are required of a number of genera of Ceratopogonini not included in this study (material not available for study) and of the relationships between the genera of the Heteromyiini, Sphaeromiini, Palpomyiini and the Stenoxenini. Finally, the higher classification of the Ceratopogonidae continues to experience changes as new taxa are discovered and our understanding of character states within the family improves. Although the analysis above provides a greater resolution of phylogenetic relationships within the Ceratopogonidae, it is clear that future investigations should test the character state distributions discussed above and rigorously analyze further characters.

Phylogenetic Relationships of Cretaceous Ceratopogonidae

In this section I summarize the phylogenetic relationships of the Cretaceous Ceratopogonidae in Canadian amber and follow with a discussion of those additional genera that have been described from other Cretaceous ambers, in the light of the cladistic analysis of the extant fauna provided above.

One of the difficulties in dealing with fossils is that often only some of the phylogenetically important characters are visible. As such, merely placing fossil lineages on a cladogram based on extant lineages may be misleading, leading the reader to believe that all synapomorphies were visible and that others were absent in the fossil. To avoid this problem, I place each fossil lineage in a separate cladogram (Figs 22A-24B) and indicate only the synapomorphies that are visible for that fossil lineage. Character states which were absent from the fossil are indicated on sister lineages. The synapomorphy numbers correspond to those in the cladogram of the extant fauna (Fig. 21), allowing for an interpretation of the fossil in the context of the phylogenetic relationships of those extant lineages.

In general, the following analysis provides an interpretation of the Cretaceous fossils at the generic level; for most, I was unable to provide further resolution of intrageneric relationships (as discussed under each genus below). The only genus which could be interpreted further was *Leptoconops*.

Each genus is treated in a phyletic sequence. As such, the earliest lineages are dealt with first and the youngest, last.

Genera present in Canadian amber.

Leptoconops. This genus forms the sister group of all other Ceratopogonidae (Fig. 21).

There have been no previous attempts to analyze the phylogenetic relationships between species of *Leptoconops*. However, the analysis of extant subgenera of *Leptoconops* above (Fig. 21) provides a basis for understanding the phylogenetic position of *L. primaevus* and other fossil members of the genus.

L. primaevus, in Canadian amber, appears to be a member of the subgenus *Leptoconops*, as indicated by the absence of stout spines on the head and legs, the number of female flagellomeres and elongate cerci, the presence of capitate sensilla on the surface of the third palpal segment, and shape of male tergite 9. However, there are no known synapomorphies for the subgenus *Leptoconops* and until such time as these are apparent, the placement of *L. primaevus* cannot be logically confirmed. Regardless, the presence of elongate cerci and capitate sensilla on the surface of the third palpal segment indicates that *L. primaevus* belongs well within the genus *Leptoconops* (Fig. 22A).

Two character states occur as homoplasies in *Leptoconops* and these are also present in *L. primaevus*: fusion of palpal segments four and five (synapomorphy 48) and the presence of katepisternal setae (synapomorphy 32). These features are discussed further under those synapomorphies above.

As indicated by Kalugina (1991), the Upper Cretaceous *L. boreus* from Taimyr amber (Russia) is also a member of the subgenus *Leptoconops* (female third palpal segment with capitate sensilla on surface, antenna with 12 flagellomeres, cerci elongate). The Baltic amber species *L. succineus* (Szadziewski, 1988) and the unnamed Sakhalin amber species of *Leptoconops* (Szadziewski, 1990a) are also members of this subgenus. Grimaldi (1993a) includes a photograph of what is clearly a female *Leptoconops* in New Jersey amber; the specimen has elongate cerci and 12 flagellomeres (palpal pit not visible in photo) and is also likely a member of the subgenus *Leptoconops*.

Based on the cladogram (Fig. 21) and the presence of Upper Cretaceous fossils of the subgenus *Leptoconops*, it is clear that the genus *Leptoconops* was present and at least partially diversified prior to the Upper Cretaceous. If the presented phylogenetic arrangement is correct, species closely related to the subgenera *Styloconops* and *Brachyconops* must have also been present during the Cretaceous and the future discovery of such fossils would not be surprising.

Protoculicoides. This monotypic fossil genus was difficult to interpret; there were not many character states available for interpretation and what was available produced two alternative hypotheses of relationship.

The first hypothesis suggests that *Protoculicoides* is the sister group to the Forcipomyiinae + *Dasyhelea* (as presented in Fig. 22B). However, the only synapomorphy placing it there is the lack of inner teeth on the female claws. This feature is difficult to observe accurately even in extant material and may have been misinterpreted in the single female of *Protoculicoides* (although a good view was available). Furthermore, the character is susceptible to homoplasy. If this hypothesis is true, the simple male claws (synapomorphy 11) are a reversal to the plesiomorphic condition. Furthermore, this phylogenetic interpretation of *Protoculicoides* suggests that synapomorphy 19 evolved before synapo-morphy 18. The analysis of *Atriculicoides* below provides the opposite interpretation (see below), with synapomorphy 18 evolving before synapomorphy 19. To resolve this conflict, it is possible that *Proto-culicoides* once had broadly contiguous eyes dorsomedially but reverted to the plesiomorphic condition, as have some extant *Dasyhelea*. The medial line of weakness present in these extant *Dasyhelea,* interpreted as a remnant of previously broadly contiguous eyes, was not visible in *Protoculicoides* and hence this idea could not be tested.

The second hypothesis of relationship suggests that the simple male claws are a good indicator of relationship. If true, *Protoculicoides* would represent the sister group to all other Ceratopogonidae other than *Leptoconops* (Fig. 22C) and synapomorphies 16 and 19 would have evolved independently in *Protoculicoides*. I consider it especially unlikely that synapomorphy 16 would have evolved twice.

Because of these uncertainties, I consider the cladistic placement of *Protoculicoides* as highly tentative, but with the first hypothesis (Fig. 22B) more likely than the second (Fig. 22C). Regardless, it does seem clear that *Protoculicoides* is an early lineage within the family.

I have tentatively put *Protoculicoides* into the subfamily Forcipomyiinae. Although this creates a paraphyletic group (i.e. Dasyheleinae arising from within Forcipomyiinae), I retain this rather unsatisfactory position for the present for two reasons. First the phylogenetic position of *Protoculicoides* is uncertain and, second, there is a need for a major restructuring of classification of the entire family at the subfamilial and tribal level based on continuing phylogenetic studies and this makes the proposal of new names for higher categories undesirable at the present time.

Boesel (1937:51) suggested that *Protoculicoides* was closely related to *Culicoides*, based on the shared presence of large humeral pits. Borkent et al. (1987:602) point out, however, that humeral pits are present in a number of early lineages of Ceratopogonidae and that this character and the distribution of its various character states are poorly known and understood, making it presently impossible to interpret cladistically.

Szadziewski (1988:28) also regarded *Protoculicoides* as a member of the Culicoidini (which he considered to otherwise include only *Culicoides, Paradasyhelea* and *Austroconops*), but the evidence for this was not clear. He further indicated that *P. depressus* was "very close" to extant *Culicoides cylindricornis* (Wirth and Blanton, 1973:430) from the Amazon Basin based on the following shared character states: no macrotrichia on the wing membrane, a long first radial cell, a high CR and the proximal flagellomeres with stout verticils. All these character states show a high degree of homoplasy and appear to be poor indicators of relationship. The stout verticils present on the basal flagellomeres identified by Szadziewski (1988) as significant appear to be similar to those present in a wide array of ceratopogonid genera. Furthermore, the elongate radial cell of *Culicoides cylindricornis* is a single cell while in *P. depressus* both radial cells are present. Finally, as discussed above under "Phylogenetic Relationships of Extant Ceratopogonidae" there are no synapomorphies which can be presently recognized to support the monophyly of the Culicoidini.

More recently, Szadziewski and Schlüter (1992) noted that *Protoculicoides* could not belong within the Culicoidini because it had no sensilla coeloconica, a feature otherwise present in the tribe (but also present in many other genera, including some *Paradasyhelea*; Borkent et al., 1987:600; see synapomorphy 27 above).

Atriculicoides. Available evidence indicates that *Atriculicoides* is an extinct sister group to Forcipomyiinae + Dasyheleinae (Fig. 22D).

It is uncertain whether *Atriculicoides* is a monophyletic group as no synapomorphies can be proposed at the present time. However, it is also apparent that all members are structurally rather similar to one another.

I am provisionally placing *Atriculicoides* in the subfamily Forcipomyiinae. As for *Protoculicoides*, this creates a paraphyletic group (i.e. Dasyheleinae arising from within Forcipomyiinae; see further discussion of this problem above under *Protoculicoides*).

When Remm (1976) first described *Atriculicoides* he noted that the genus shared features of both *Atrichopogon* and *Culicoides* but he did not analyze its phylogenetic position. Szadziewski (1988) indicated that

Atriculicoides was a member of the tribe Culicoidini but the reasons for this conclusion were unclear. Szadziewski and Schlüter (1992) state that the genus could not be a member of the Culicoidini because it lacks sensilla coeloconica (but see synapomorphy 27 and discussion under *Protoculicoides* above) and the males have four elongate terminal flagellomeres (*Culicoides* and *Paradasyhelea* have three). The character state distribution of four elongate flagellomeres has not been described but may be a synapomorphy of *Atriculicoides*, Forcipomyiinae + Dasyheleinae. Further study is required.

The character state distribution of sensilla coeloconica on the first flagellomere has never been explicitly analyzed. It is present in a number of genera in the Ceratopogoninae (Wirth and Grogan, 1988) and may well be a valid synapomorphy for this subfamily (but with numerous subsequent losses). Further analysis is required (see synapomorphy 27 above).

Adelohelea. There is some uncertainty about the exact phylogenetic position of this genus (Fig. 23A). The genus is circumscribed by synapomorphies 15, 16, and 28'. As such, it could be most closely related the Forcipomyiinae + *Dasyhelea, Washingtonhelea, Culicoides, Paradasyhelea*, the ancestor of the remaining Ceratopogoninae or a combination of some of these.

I provisionally consider the genus a member of the Ceratopogoninae.

Heleageron. Although this new genus can be differentiated from other described genera, its specific phylogenetic position is also somewhat unclear, at least partially because some character states were not visible. Those that could be seen, show that the genus is circumscribed by synapomorphies 16 and 28' (Fig. 23B). As such, *Heleageron* may be most closely related to the Forcipomyiinae + *Dasyhelea, Washingtonhelea, Culicoides, Paradasyhelea*, the ancestor of the remaining Ceratopogoninae or a combination of some of these.

I provisionally consider the genus a member of the Ceratopogoninae.

Culicoides. The phylogenetic position of this extant genus is near the very base of the Ceratopogoninae (Fig. 21).

Because of overall morphological similarity and a lack of any conflicting synapomorphies (which might suggest that some of the species belong elsewhere), I am treating all Canadian amber *Culicoides* as a single group. However, there is no cladistic evidence that they are monophyletic. Furthermore, there is a question as to whether the species actually belong to the genus. I have compiled all information on the Cretaceous *Culicoides* species as a group and show their phylogenetic position in Figure 23C. Readers wanting further details may compare the specific descriptions with character states on the cladogram.

Extant *Culicoides* form a monophyletic group based on the presence of at least three flagellomeres with sensilla coeloconica (see synapomorphy 27) (Fig. 21). Of those fossil species where the female are known and the antennae are clearly visible, *C. filipalpis, C. tyrrelli*, and *C. bullus* all have sensilla coeloconica restricted to the first flagellomere. Combined with other synapomorphies, this suggests that these species may be most closely related to *Culicoides, Paradasyhelea*, the ancestor of the remaining Ceratopogoninae (defined by synapomorphy 28') or any combination of these unresolved lineages (Fig. 23C).

This hypothesis of relationship indicates that the Cretaceous species of *Culicoides* might be recognized as a separate genus but there are four reasons why I have not done so. First, I consider the single synapomorphy grouping extant *Culicoides* (synapomorphy 27) to be only a weak indicator of relationship. As such, it would be preferable for further evidence of monophyly to be acquired before the classification is modified. Second, I was unable to find any synapomorphies grouping the Cretaceous *Culicoides*; their recognition as a new genus would create another taxon of uncertain monophyly. Third, the presence of sensilla coeloconica is difficult to detect, making it impossible to classify those species for which either females are unknown or where the antennae cannot be observed in detail if a new genus was recognized. Fourth, the fossil *Culicoides* are otherwise similar to extant *Culicoides* in many details and overall habitus (and therefore recognizable). In addition, most fossil *Culicoides* in Eocene-Lower Oligocene Baltic amber appear to have sensilla coeloconica restricted to the first flagellomere (Szadziewski, 1988); only one species, *C. balticus* Szadziewski, also has sensilla coeloconica present on flagellomere 5. Otherwise these Eocene-Lower Oligocene species, too, are similar to extant members of *Culicoides*. Their lack of more extensively distributed sensilla coeloconica would necessitate placing them in a new genus with the Cretaceous *Culicoides* and this would likely be premature.

I was unable to detect any synapomorphies which might suggest specific phylogenetic relationships between the species of fossil *Culicoides*.

There is some question of the generic placement of *C. casei* from New Jersey amber, as discussed by Grogan and Szadziewski (1988). Although they tentatively placed the species in *Culicoides*, they also indicated that some features were similar to those of extant *Ceratopogon*. The uncertainty of the state of the palisade setae, and the fact that three synapomorphies of *Ceratopogon* could not be seen in *C. casei* (see "Phylogenetic relationships of extant Ceratopogonidae", synapomorphies 29-31) meant that this possibility could not be tested any further.

Minyohelea. Although many character states were not visible, those that were indicated that *M. pumilis* belongs within that group of genera now considered to be members of the Ceratopogonini (Fig. 23D).

Male *M. pumilis* lack apicolateral processes and hence exhibit the plesiomorphic condition of synapomorphy 3. However, as discussed under that character, the apomorphic state is clearly susceptible to reversals. Considering the presence of other synapomorphies in *M. pumilis*, it is likely it too secondarily reverted to the plesiomorphic state.

There no further data which would conclusively indicate where *Minyohelea* belongs within that group of genera delineated by synapomorphies 28" and 42. There are some similarities with the extant genus *Camptopterohelea* which bear mentioning. Both genera have a reduced palpus, flagellomere 1 lacks sensilla coeloconica, the females have antennal flagellomeres of nearly equal length and have very short mouthparts, and the wing venation is reduced, with a single radial cell. However, in *Camptopterohelea* the female has widely spaced eyes dorsomedially, the males lack an antennal plume, the legs are moderately stout, the claws are single and palisade setae are also present on the second tarsomere of the hindleg.

Minyohelea is also somewhat similar to the lineage *Baeohelea* + *Baeodasymyia* in sharing a reduced wing venation (also present in several other taxa) and a reduced maxillary palpus (also susceptible to homoplasy; see synapomorphies 37 and 48).

Future phylogenetic investigations of extant genera are required before further resolution of the position of *Minyohelea* might be possible. In addition, the discovery of future specimens of *Minyohelea* may allow for further character states to be observed and interpreted.

Palaeobrachypogon. This genus belongs within a large group of poorly understood genera defined by the presence of synapomorphy 28" and absence of synapomorphy 42 (Fig. 24A). Although members of *Palaeobrachypogon* are similar to extant *Brachypogon* in overall habitus, there are no shared synapomorphies between the two providing evidence of close phylogenetic relationship.

Peronehelea. This genus belongs within a large group of poorly understood genera defined by the presence of synapomorphy 28" and absence of synapomorphy 42 (Fig. 24B).

The presence of a single, elongate, female hindleg claw is unique within the Cretaceous ceratopogonid fauna. Based on outgroup comparisons with other families of Nematocera, such an elongate hindclaw is unique and hence should be interpreted as a synapomorphy. However, although the single, elongate claw is restricted to that group defined by synapomorphy 28" it appears to be homoplastic, present in a number of genera originating before synapomorphy 42, in some

Serromyia Meigen, *Schizohelea* Kieffer, some Heteromyiini and some Sphaeromiini. Furthermore, it is apparent that the feature can be lost within a given genus (Borkent and Bissett, 1990). As such, this homoplasy makes it presently impossible to use this derived character state as evidence of close relationship.

Other Genera recorded from Cretaceous deposits.

Austroconops. A species of this genus was described by Szadziewski and Schlüter (1992) from a single female in French amber. Although the single synapomorphy identified for *Austroconops* by Borkent et al. (1987) (described above as synapomorphy 14) could not be seen in the specimen, the relative lengths of the antennal flagellomeres, wing venation and microtrichia, reduced palpus, and small claws all suggest the fossil is correctly identified. Furthermore, the phylogenetic position of the genus as an early lineage within the family combined with the presence of more derived lineages recorded from Cretaceous deposits indicates that this genus or an immediate ancestor was present at that time.

Szadziewski (1990b) noted the possibility that the Lower Cretaceous specimen from Australia identified by Jell and Duncan (1986:179) as 'simuliid? indet. 2' was actually a member of *Austroconops*. However, further observations are required before this can be confirmed.

Leptohelea. Remm (1976) originally placed the single known male fossil from Taimyr amber in the genus *Baeohelea*. However, Wirth and Grogan (1988:24) considered that it likely belonged to a different, extinct, unnamed genus and Szadziewski (1988:50) suggested that a number of characters indicated that the species belonged in *Leptohelea*, a recent genus in which only the female is known.

The following character states were suggested by either Szadziewski (1988) or Wirth and Grogan (1988) as indicating that *L. taimyrica* could not be placed in *Baeohelea*. Each character state is identified and is followed by a discussion of its phylogenetic significance. It is important to note that *Baeohelea* and *Baeodasymyia* are sister taxa and are very similar structurally (Szadziewski and Grogan, 1994a).

- male antenna with nine flagellomeres. The fusion of all flagellomeres except for the last, as shown by Remm (1976:Fig. 5b), suggests that *L. taimyrica* actually belongs to that group of genera in which flagellomeres 1-11 are fused (synapomorphy 33', see Fig. 21) and which was discussed by Grogan and Borkent (1992). It would be invaluable to examine the fossil to see if flagellomere 11 is actually involved in the fusion.
- a single radial cell. Both *Baeohelea* and *Baeodasymyia* have lost the radial cells altogether. The presence of one radial cell in *L. taimyrica* may be part of a morphocline leading to the complete loss but this condition occurs in a number of other genera as well.

- radial cell elongate. The figure of the wing of *L. taimyrica* shows a costal ratio of about 0.55. Male *Baeohelea nana* Wirth and Blanton have a ratio of 0.40-0.44 (pers. obs.) and I consider these differences to be slight and within the range of intrageneric variation of some other extant genera of ceratopogonids.
- cell r_{4+5} with intercalary veins. This is a puzzling feature of the wings of many ceratopogonids and may be found in some taxa of early lineages (e.g. Forcipomyiinae) as well as in relatively young lineages (e.g. *Echinohelea*). The character state distribution has never been analyzed but their appearance sporadically in a wide number of taxa suggests the presence of homoplasy. Regardless, there is no evidence that this feature is presently of value in distinguishing genera or relationships.
- apicolateral processes. Although *L. taimyrica* has well developed apicolateral processes, *Baeodasymyia* also has short apicolateral processes with each bearing a single apical seta. Differences in the lengths of apicolateral processes are a feature commonly varying within extant genera and cannot be considered as evidence of relationship (or lack thereof), at least at this level of analysis.

The presence of a 2 segmented palpus in *L. taimyrica* is a character state here regarded as a synapomorphy of *Baeohelea* + *Baedasymyia* (Szadziewski and Grogan, 1994a), indicating the *L. taimyrica* may have been placed corrected by Remm (1976) in *Baeohelea*. Alternatively, the fused antennal flagellomeres indicate relationship with *Brachypogon* and its immediate relatives. Either way, available evidence does not support the placement of the fossil in *Leptohelea*. However, I am reticent to suggest another generic reallocation without examining the fossil firsthand.

Finally, it is important to note that some other Ceratopogonidae also have a reduced number of palpal segments and it may be that *L. taimyrica* has more than the two reported (these can be difficult to observe accurately), in which case the available evidence would indicate that it is actually a member of, or a sister group to, the extant genus *Nannohelea* which has 3 palpal segments and fused male flagellomeres including flagellomere 11.

Conclusions

There is one very obvious pattern evident from this cladistic analysis of Cretaceous Ceratopogonidae: all taxa form very early or relatively early lineages within the family (Fig. 25). None exhibit synapomorphy 42 or synapomorphies of taxa subsequently derived from the lineage defined by synapomorphy 42 (see Fig. 21). As such, this interpretation of the fossils is a confirmation of the cladogram determined for extant taxa (i.e. early lineages determined by cladistic analysis are those represented by Cretaceous fossils). Some early lineages identified by the cladistic analysis of extant taxa are represented only by extinct Cretaceous genera (i.e. *Atriculicoides*) while for others the genera have existed since at least the Upper Cretaceous (i.e. *Leptoconops, Austroconops,* possibly *Culicoides)* (Fig. 25).

Szadziewski (1988:243) considered members of each of the Heteromyiini, Sphaeromiini, Palpomyiini and Stenoxenini to be old lineages and the absence of Sphaeromiini and Stenoxenini in Baltic amber to be due to their presence in the southern hemisphere during the Eocene. The analysis here suggests that they may have been absent during the Eocene because they are young groups. There is no evidence of any lineage with synapomorphy 42 (Figs 21, 25) (which includes the Heteromyiini, Sphaeromiini, Palpomyiini, and Stenoxenini) in Cretaceous amber. These patterns are discussed further below under "Comparisons of Mesozoic ceratopogonid fossil assemblages".

DISCUSSION

Palaeoecology

Russell (1977, 1989) described the environment present in Alberta during the Cretaceous and this provides a context in which Canadian amber ceratopogonids can be best understood.

Dinosaur Provincial Park in southern Alberta is one of the best known Upper Cretaceous fossil sites in the world and is located only 110 km directly north of the site at 5 km south of Grassy Lake (Fig. 20B). The general environments of these two sites were likely very similar and previous studies of this famous park give special insight into the environment in which the amber from 5 km south of Grassy Lake was originally formed.

The deposits at Dinosaur Provincial Park are dated at 75 million years and represent a meandering river system which flowed from the southwest in Montana and in the park itself flowed in an easterly direction. Cordilleria (proto Rocky Mountains) lay 320 km to the west but apparently produced no rainshadow effect in the park. To the east the midcontinental ocean was 100 km away (Fig. 20A) but the area between Dinosaur Provincial Park and the ocean was so flat that the park experienced some tidal incursions. The rivers which flowed through the park averaged about 50 meters wide and four meters deep but some larger channels were a third of a kilometer wide and 15 meters deep. These rivers deposited large amounts of sand along their banks and sandbars and small beaches were located along their lengths. Mudflats were also present along portions of these rivers. Adjacent areas bordering the rivers were extensive floodplains which left rich soils further enriched by volcanic ashes and supported a lush vegetation. Analysis of fossil spores and pollens indicates that a wide array of plants were in the park. Water lilies floated on ponds and oxbow lakes, cattails formed extensive marshes in places and horsetails were present in wet areas. Relatives of bald cypress were common and likely formed tall, dark forests in the uplands between the rivers. A variety of conifers are recorded from southern Alberta and included the species of Araucariaceae (possibly *Agathis*) which was the likely source of Canadian amber. *Alnus, Platanus* and taxa related to mistletoes, cycads, tree ferns, grape and lilies were also present (Bell, 1965; Jarzen, 1982). Of the 85 identified plants in Dinosaur Provincial Park, angiosperms are most abundant and indicate a subtropical to tropical climate. Fossil conifer wood is common in the park and suggests that these trees formed an important component of the flora. Some logs were up to 60 centimeters in diameter and over 17 meters in length and indicate that substantial forests were likely present in places. These plants and the presence of epiphytes also confirm that the local climate was warm and humid.

The region experienced a seasonality based on precipitation as indicated by the presence of tree rings (Wolfe and Upchurch, 1987), the shed leaves of bald cypresses, and growth rings in the bones of champsosaurs (crocodile-like reptiles). Most plants, however, kept their leaves and the mean annual temperature was about 20°C with an average range of about 16°C-24°C (Wolfe and Upchurch, 1987). The annual rainfall was likely about 1000-1400 mm per year but varied somewhat seasonally and from year to year (Wolfe and Upchurch, 1987). Flecks of charcoal indicate that the region experienced enough aridity to allow for the presence of periodic forest fires.

Topographically, in many ways the area in Dinosaur Provincial Park 75 million years ago was similar to the Mississippi Delta in southern Louisiana, U.S.A. today. The climate, however, was far more equable and the silt load in the rivers was much less due to a much smaller and more precipitous drainage area (which likely influenced the quality of many aquatic habitats).

The park had a rich vertebrate fauna, with nearly three dozen types of dinosaurs including hadrosaurs (duckbills), ceratopsians, tyrannosaurids, and small theropods. The most common of the large dinosaurs (2-3 metric tonnes) were the duckbilled dinosaurs and these likely frequented the margins of aquatic habitats. Russell (1989:142-143) provides a delightful color illustration of hadrosaurs roaming a river bank in southern Alberta and this probably provides a very reasonable facsimile of the general environment in which many of the Ceratopogonidae now preserved in Canadian Cretaceous amber previously lived.

Of the more than 30 types of dinosaurs, at least 11 weighed less than 150 kg. Other vertebrates included pterosaurs, turtles, crocodiles, salamanders, frogs, and a number of small (about the size of marmots or smaller) mammals. Although not recorded from the park itself, fully developed birds were present elsewhere in North American during the Upper Cretaceous (Romer, 1966:374), including Gaviiformes (loons), Podicipediformes (grebes), Ciconiiformes (including herons, flamingos), and Charadriiformes (gulls, terns, auks). Pelecaniformes (pelicans) are known from the Cretaceous of Europe and the Palaeocene of North America.

Hesperornithiforms, flightless, loon-like marine birds and *Ichthyornis*, a gull-like toothed bird, are also known from the Upper Cretaceous of Kansas at 80 million years (Russell, 1989) and the order clearly has a much older history of diversification (Sereno and Chenggang, 1992).

The presence of large numbers of dinosaurs undoubtedly had a major impact on some habitats. Their trampling along stream and river margins likely left extensive mudholes and mudflats. In addition, dinosaur dung production was high and has been estimated at 500 metric tonnes/ km^2/year. Some extant species of *Culicoides* breed in the dung of large herbivores such as elephants (Meiswinkel, 1992) and some Cretaceous species of *Culicoides* may have done the same.

The debate about whether the dinosaurs were warm-blooded has not been completely resolved (Bakker, 1972; Béland and Russell, 1979; Marx, 1978; Thomas and Bland, 1980). Nevertheless, it seems likely that at least some were at least incipiently warm blooded (Russell, 1989).

The deposits at Choteau, Montana, about 220 km due south of the site at 5 km south of Grassy Lake (Fig. 20B), also provide some information about the type of environment in which Upper Cretaceous Ceratopogonidae lived. The site is also dated at 75 million years, represents a semi-arid environment with a well defined seasonality in precipitation, and was at a somewhat higher elevation than Dinosaur Provincial Park. Here lush vegetation also grew for at least part of the year. As for Dinosaur Provincial Park, this site had rivers and supported a large population of breeding hadrosaurs (extensive nesting sites present), at least on a seasonal basis (evidence indicates that the Choteau hadrosaurs exhibited migratory behaviour).

Some palaeoecological information is available from the site of deposition at 5 km south of Grassy Lake, Alberta. The following have been found associated with the coal seam bearing the amber (Heie and Pike, 1992): freshwater algae (*Pediastrum*), intact spore sacs of mosses and ferns, pollen from a variety of gymnosperms, wood fragments and sections of tree trunks and a wide array of vertebrates (champsosaurs, crocodiles, soft-shelled turtles, hadrosaurs and tyrannosaurs).

The presence of Cretaceous Ceratopogonidae indicates that a diverse array of aquatic and semiaquatic habitats were present in southern Alberta. *Leptoconops* larvae generally live in wet sand and the identification of a number of specimens of this genus (Table 1) suggests that such habitat was abundant (as confirmed by extensive sandstone deposits in the general area). Although other extant ceratopogonids are found in many different kinds of aquatic habitats including rivers and lakes, early extant lineages are restricted to smaller aquatic, semiaquatic or merely wet environments such as bogs, fens, marshes, tree holes, seeps, wet and rotting vegetation and manure. It therefore seems likely that Cretaceous ceratopogonids (other than *Leptoconops*), as early lineages, were present in similar sorts of habitats in southern Alberta.

As hypothesized below under "Evolution of blood-sucking and host selection in Ceratopogonidae", female ceratopogonids of both vertebrate and invertebrate blood-sucking types were present. At least two, and possibly three, species likely took blood from dinosaurs (see below).

Szadziewski (1988:265) reported that males comprised 37.4% of the

Ceratopogonidae in Baltic amber. Saxonian amber included 44.0% males (Szadziewski, 1993; calculated from his table 1). Szadziewski (1988) suggested that this was similar to the ratio obtained by sampling the extant fauna with yellow pan traps in Poland (41% males) and emergence traps in Germany (31% males). Further information corroborates this interpretation; Knausenberger (1987) found 43.7% males in his rearings of 5,535 ceratopogonids in the eastern United States (calculated from his Table 6). The pattern in Canadian amber is the same, with males making up 43.7% of the fauna (Table 1). The similarity between the fossil percentages and those from the extant fauna likely indicates that overall, the amber was non-selective in the preservation of male and female Ceratopogonidae and that the amber therefore probably preserved the ceratopogonids near the site of their emergence. As male ceratopogonids generally remain near the site of emergence while females disperse (Blackwell, et al., 1992a; Bidlingmayer, 1961; Linley and Davies, 1971:268; Zimmerman and Turner, 1984), it would be expected that the male/female ratio would decrease with increased distance from the site of emergence.

The low percentage of males in the lumped emergence data above is different from that reported for some groups of *Culicoides* species. Based on reared larvae, Kettle (1955) found 52.2% of 31 species were females. Cameron (1948) reported 47.8% females for a group of 12 species collected with emergence traps. However, Cameron (1947) found 34.9% males in a sample including 8 species of *Culicoides*. Clearly, more studies are needed on the sex ratio of emerging ceratopogonids.

The hypothesis that Canadian amber Ceratopogonidae were preserved near the site of their emergence is further supported by the observation that both Ceratopogonidae and aphids, which were also intimately associated with the amber producing trees (Heie and Pike, 1992), appear to be among the most common groups in the amber (McAlpine and Martin, 1969; Pike, 1994).

Of the 58 specimens Remm (1976) studied in Taimyr amber, 22.4% were males (as calculated from his listing of material). This low number of males may indicate that the preserving amber was some distance from the emergence site of Taimyr ceratopogonids. However, two factors may have influenced this apparently low number. First, one piece of amber included 18 females of one species. If this is considered as a single record, the male ratio would be 31.7%. Second, Remm (1976) noted that there were a total of 101 specimens originally available for study but some were unsuitable for study. He did not report the sex of those specimens and this too, may have biased the relative number of males and females.

Canadian amber from Grassy Lake and Cedar Lake contains a remarkable diversity of arthropods, including the following (n= number of families recorded): Tardigrada (1), Araneida (4), Acarina (2), Pseudoscorpiones (1), Collembola (1), Blattodea (1), Dermaptera (1), Mantodea (1), Psocoptera (1), Thysanoptera (1), Hemiptera (1), Homoptera (4), Neuroptera (1), Hymenoptera (14), Coleoptera (3), Trichoptera (1), Lepidoptera (1) and Diptera (18) (Beardsley, 1969; Botosaneanu and Wichard, 1983; Brown and Pike, 1990; Cooper, 1964; Gagné, 1977; Hamilton, 1971; Heie and Pike, 1992; Klimaszewski and Kevan, 1986; Krzeminski and Teskey, 1987; MacKay, 1970; McAlpine, 1973; McAlpine and Martin, 1966; Peterson, 1975; Poinar, et al., 1993; Richards, 1966; Schawaller, 1991; Teskey, 1971; Wilson, 1985; Yoshimoto, 1975).

Poinar et al. (1993) showed that some mites in Canadian amber were parasitic on Ceratopogonidae (as reported earlier by Ewing, 1937). The ceratopogonid shown in their photograph is likely a member of *Atriculicoides globosus* (see under that species above). The mite larvae noted in my study were also likely parasitic on the associated Ceratopogonidae, although none had their mouthparts attached. Parasitic mites have also been recorded from ceratopogonids in Lebanese amber (Grimaldi, 1993a, in photo; Poinar et al., 1994), indicating that the parasitic relationship between mites and Ceratopogonidae (Wirth, 1977) has existed since at least the Lower Cretaceous (Hauterivian).

The fossil assemblage recorded from Grassy Lake is consistent with that hypothesized as characteristic of the Upper Cretaceous environment at Dinosaur Provincial Park. Heie and Pike (1992) indicated that angiosperm pollen was rare at the amber site and this is strikingly inconsistent with reports of abundant angiosperms at other nearby locations (Jarzen, 1982; Norris, 1975; Retallack and Dilcher, 1986; Wolfe and Upchurch, 1986). In addition, the suggestion by Heie and Pike (1992) that "winters were considerably milder than now" is misleading. All available evidence strongly suggests that the area had no winter (i.e. freezing temperatures) at all.

The palaeolatitude of the Grassy Lake site during the Campanian was about 50°N (Smith et al., 1981), confirming the conclusion that the area experienced a warm, mesothermal environment (Wolfe and Upchurch, 1986).

The presence of Grassy Lake amber within a coal deposit suggests that the amber captured its insect inclusions at a time of lush vegetation. The formation of extensive organic deposition allowing for the formation of this coal indicates the presence of extensive swamps, likely similar to those that formed at nearby Drumheller at about 71 million years ago (Russell, 1977), about 200 km NW of Grassy Lake (Fig. 20B). The deposits at Drumheller indicate the presence of extensive bald cypress swamps in which the trees bore epiphytes. Peatbogs were present along the edges of the delta. Better drained uplands in the area supported a variety of Taxodiaceae including redwoods, china firs, torreyes and sequoias. These tall trees likely formed a dense overstory and resulted in a dark forest floor with abundant mosses and ferns. On raised levees grew sun-loving ferns, cycad like plants, ginkgo, magnolias, gunneras and Katsura trees (broad-leafed). On poorly drained soils were ash, swamp hickory, swamp oak, box, myrtle, sweetleaf, protea and members of the poison-ivy family. These sorts of habitats are likely very similar to those in which the Grassy Lake and Cedar Lake Ceratopogonidae were preserved.

At about 70 million years ago major changes occurred in Alberta. A large delta formed on the eastern land margin, the sea eventually retreated 900 km to the east by the end of the Cretaceous at 65 million years and temperatures increased by about 3°C. These changes must have had an important impact on at least the distribution of the Ceratopogonidae living in southern Alberta.

Comparisons of Mesozoic ceratopogonid fossil assemblages.

Aside from those in Canadian amber, Mesozoic Ceratopogonidae have been noted from six other deposits, all of which are Cretaceous in age (see Table 2 for specific dating).

Grogan and Szadziewski (1988) described a *Culicoides* from New Jersey amber (based on a single male). This amber was described in detail by Grimaldi et al. (1989).

Langenheim et al. (1960) reported an unidentified ceratopogonid (as Heleidae) from Alaska. I attempted to locate and study the specimen but with no success. The specific age of this Cretaceous amber is uncertain.

The Ceratopogonidae from the Upper Cretaceous of the Taimyr Peninsula in Russia were described by Remm (1976). He recognized nine species (based on 58 specimens) and incorporated two of these into a new genus *Atriculicoides* Remm (Table 18). Recently Kalugina (1991) has described a species of *Leptoconops* from the same deposits. The Taimyr amber Ceratopogonidae are presently being restudied by Szadziewski (in prep.).

The Taimyr amber was collected from the right bank of the Maimechi River about 5-6 km above its mouth in the Yantardakh area of the Eastern Taimyr (Remm, 1976) and was identified as originating from the Coniacian-Santonian in the Kheta Formation by Zherikhin and Sukatsheva (1973).

Szadziewski and Schlüter (1992) described three species from the Upper Cretaceous (Cenomanian) of France (Table 18) and noted, in addition, the presence of a female *Leptoconops* previously recorded by Schlüter (1978). The specimens in amber were found in clay or silts near Ecommoy, France (Schlüter, 1983). Two of the species named by Szadziewski and Schlüter (1992) are members of *Atriculicoides* and are discussed above under the treatment of that genus. The other two species present in the amber are members of *Leptoconops* and *Austroconops*, both representing the earliest two lineages, respectively, within the phylogeny of extant Ceratopogonidae (Borkent et al., 1987).

Ceratopogonidae have been reported from Lower Cretaceous ambers from Lebanon (Grimaldi, 1993a; Poinar et al., 1994; Schlee and Dietrich, 1970; Schlee and Glöckner, 1978:62) and Austria (Schlee, 1984:32). These specimens have not yet been described but will certainly provide fascinating information on the diversification of the family when they are properly analyzed.

Other possible Mesozoic members of the Ceratopogonidae are discussed above in "Questionable fossil Ceratopogonidae". Grimaldi (1990:179) provided a list of Cretaceous Ceratopogonidae but failed to include species described by Remm (1976).

Even though only a limited number of specimens are known from the above Cretaceous ambers (Table 2), some patterns are evident. Both *Leptoconops* and *Atriculicoides* are present in Canadian, Taimyr and French amber indicating that these two genera were widespread. *Culicoides* are known from Canadian, New Jersey and Taimyr amber and *Palaeobrachypogon* is present in both Canadian and Taimyr amber. This suggests that these groups of ceratopogonids were also widespread in at least the northern hemisphere.

The ceratopogonid faunas in Canadian and Taimyr ambers (the two with the most specimens) are strikingly similar and share at least four genera and one species (*Culicoides filipalpis*) (further analysis by R. Szadziewski (in prep.) of the Taimyr ceratopogonids will likely show further affinities). This possibly reflects similarities in environmental conditions in both southern Alberta and northern Russia at that time, although the Taimyr Peninsula was almost certainly much cooler than southern Alberta (Barron, 1987; Wolfe and Upchurch, 1987). Most available evidence indicates the presence of a continuous Upper Cretaceous land connection between western North America and the eastern Palaearctic, including the Taimyr Peninsula (Dixon, 1989:9). Marincovich et al. (1990), however, suggest a narrow Beringian gap at this time. Such a direct land connection, or at least closely approximating land, with continuous forest is further supported by the shared ceratopogonid fauna in Canadian and Taimyr amber reported here. The recent discovery of an Upper Cretaceous vertebrate fauna (including dinosaurs, pterosaurs and mammals) and of megafloral and palynological specimens indicating a diverse deciduous coniferous flora from northern Alaska, which was at a similar latitude as the Taimyr Peninsula, supports the hypothesis of significant biotic connections between western

North America and the northern portion of the eastern Palaearctic (Brouwers et al., 1987; Davies, 1987; Spicer and Parrish, 1987; Wolfe and Upchurch, 1987). Nevertheless, it is important to recognize the ecological variability present in extant genera of Ceratopogonidae and the potential of at least some lineages for dispersal (Borkent, 1991).

Patterns of diversification, species turnover and extinction in Ceratopogonidae and other insects.

Rapid progress has been made during the past six years of our understanding of the fossil record of Ceratopogonidae, mostly through the efforts of R. Szadziewski in Poland. In particular, his impressive study of Eocene-Lower Oligocene Baltic amber ceratopogonids (Szadziewski, 1988) allows for important comparisons with the Upper Cretaceous diversity reported here. Further detailed information is available from 22 million year old, Miocene Saxonian amber (Szadziewski, 1993) and from Dominican Republic amber (aged 20-30 million years, Davis (1989); Szadziewski and Grogan (1994a, 1994b)).

To make comparisons between the diversity of Upper Cretaceous and Eocene-Lower Oligocene Baltic amber ceratopogonids, the fossil taxa were incorporated into a cladogram of extant taxa (Figs 25 and 26, respectively). The diversity present in Saxonian amber (see below) is similar to that in Baltic amber. In addition, amber from the Dominican Republic includes only extant genera (see below), including a number of derived lineages (Szadziewski and Grogan, 1994a, 1994b).

The cladogram incorporating all Cretaceous fossils (Fig. 25) shows two important features. First, all recorded taxa represent lineages which diverged before the group defined by synapomorphy 42 and therefore are relatively early lineages within the family. Second, although some Cretaceous fossils are known as extant genera (*Leptoconops, Austroconops, Culicoides* (but see "Phylogenetic relationships of Cretaceous Ceratopogonidae" above regarding *Culicoides*), some of the early lineages are represented by extinct genera. In particular, the lineage represented by the extant group Forcipomyiinae + *Dasyhelea*, is known in the Cretaceous only as the extinct genus *Atriculicoides* and possibly, *Protoculicoides*. As noted below, these extinct genera have no identifiable autapomorphies and therefore may actually form ancestral groups; as such, they may not be strictly speaking extinct.

It is highly likely that the family has a history which predates the Cretaceous. It is striking that the subgenus *Leptoconops* is present in Upper Cretaceous amber, indicating that the genus itself must have diversified into extant subgenera at an earlier time. Similarly, the occurrence of Upper Cretaceous taxa with synapomorphy 28" (*Palaeobrachypogon* and *Peronehelea*) indicates that more basal cladogenesis occurred at an earlier time. It certainly seems realistic to suggest that the family arose and started to diversify during or before the Late Jurassic. The presence of unidentified fossil ceratopogonids in Lower Cretaceous Lebanese amber aged 127-130 million years (Hauterivian) (Grimaldi, 1993a) and in Austrian amber (Schlee, 1984) dated at 120-130 million years also supports this suggestion. Finally, the presence of Chironomidae, as the sister group of the Ceratopogonidae, in Jurassic deposits (Kalugina, 1985) argues for the presence of Ceratopogonidae during that period.

Although figure 25 shows the occurrence of all Upper Cretaceous ceratopogonids on the cladogram, there are indications of a further pattern. Most of the fossils shown are present in Canadian amber, dated at about 71-84 million years ago; only *Austroconops* was not present. Taimyr amber (83-89 million years) includes *Atriculicoides, Culicoides* and *Palaeobrachypogon* (*Leptohelea taimyrica* is of uncertain but most likely more derived affinity). The older French amber, dated at about 93-99 million years, is known only from a few specimens (Table 2) and includes only three genera of Ceratopogonidae: *Leptoconops, Austroconops*, and *Atriculicoides*. These represent only the earliest lineages of Ceratopogonidae, all diversifying before the appearance of synapomorphies 17" and 26. This provides initial evidence that the group defined by synapomorphies 17" and 26 diversified between 83 and 99 million years ago.

Culicoides casei, dated at 93-99 million years in New Jersey amber, may be an exception to this interpretation because the species may indicate that the genus *Culicoides* predates this period (and hence the node defined by synapomorphies 17" and 26. However, there is some uncertainty regarding critical character states of this species and of its true generic placement. Available information indicates that it may represent a lineage predating synapomorphies 17" and 26. Further study (or material) is required.

The ceratopogonids in Baltic amber show a significantly different pattern (Fig. 26) from those from the Upper Cretaceous (Fig. 25). First, the fossils represent lineages occurring throughout the phylogeny of the Ceratopogonidae, including representatives of extant genera in the Heteromyiini (*Physohelea*) and Palpomyiini (*Palpomyia, Bezzia*). The presence of Eocene-Lower Oligocene fossils of these two tribes shows that all earlier nodes in the cladogram predate this epoch.

Second, all fossils of early lineages of Ceratopogonidae are represented as extant genera (*Leptoconops, Forcipomyia, Atrichopogon, Dasyhelea, Culicoides, Ceratopogon*). None of these early lineages are of extinct genera, indicating that these early groups had differentiated into extant genera by the Eocene-Lower Oligocene.

The only extinct genera recorded in Baltic amber are those delineated by synapomorphies 28" and 42 and which otherwise incorporates a number of phylogenetically poorly understood extant lineages (Figs 21,26). It is possible that some fossil genera may yet be discovered as extant taxa, as has been the case with the extant Australian genus *Chimaerohelea* Debenham (1988) which was also described by Szadziewski (1988) from Baltic amber as the apparently extinct genus *Meunierohelea* Szadziewski (see Szadziewski, 1993). If the fossil genera are considered as valid taxa, their phylogenetic position demonstrates only the possibility of limited cladogenesis at the generic level since the Eocene. Although some genera may have arisen since the Eocene, all major lineages had already diversified before then. More detailed phylogenetic studies of all taxa delineated by synapomorphies 28" and 42 are required before the significance of these fossils can be interpreted any further.

Saxonian amber, dated at 22 million years, includes only taxa that are also found in Baltic amber (Szadziewski, 1993): *Forcipomyia, Dasyhelea, Culicoides, Ceratopogon, Brachypogon, Nannohelea, Stilobezzia, Meunierohelea, Monohelea, Serromyia, Fossihelea* Szadziewski, *Mantohelea* Szadziewski, *Eohelea* Petrunkevitch, and *Palpomyia*. All but three of these are extant genera (Fig. 26). The presence of these taxa confirms the picture provided by Baltic amber ceratopogonids: most cladogenesis at the generic level occurred before this time. Similarly, amber from the Dominican Republic, dated at 20-30 million years (Davis, 1989), includes only extant genera (Szadziewski and Grogan, 1994a, 1994b): *Forcipomyia, Atrichopogon, Dasyhelea, Culicoides, Brachypogon, Nannohelea, Baeodasymyia, Stilobezzia, Heteromyia, Palpomyia*, and *Phaenobezzia*. The latter three genera are members of the most derived taxa (Heteromyiini and Palpomyiini) (Fig. 21). The presence of these genera also confirms the conclusion that most generic differentiation in Ceratopogonidae occurred prior to the Eocene.

Comparison of the phylogenetic relationships of Ceratopogonidae in Upper Cretaceous (Fig. 25) and Eocene-Lower Oligocene (Fig. 26) amber provides an interesting outline of the diversification of this family. Only early lineages are present in the Upper Cretaceous and by the Eocene-Lower Oligocene, all major lineages of Ceratopogonidae were present. This pattern shows that between 34-55 and 71-84 million years ago there was marked cladogenesis within the group defined by synapomorphy 28" and that all early lineages had differentiated into extant genera.

Furthermore, the presence of members of Sphaeromiini and Palpomyiini in Eocene-Lower Oligocene amber combined with the absence of any taxa from the Upper Cretaceous beyond synapomorphy 42, indicates that those lineages between synapomorphies 42 and 52",53" likely evolved during a period encompassed by the very latest Upper Cretaceous (Maastrichtian) and the Eocene.

It is noticeable that the tribes Sphaeromiini and Stenoxenini are not present in any of the fossil assemblages. Szadziewski (1988:252) suggested that these two tribes might be of Gondwanian origin and therefore not present in Baltic amber. Sphaeromiini have a worldwide distribution (including in the Holarctic) and the Stenoxenini are pantropical, occurring as far north as the southern United States. However, considering their phylogenetic position (Fig. 21), their absence from Baltic (Fig. 26) and Saxonian amber may equally well be the result of recent diversification. Alternately, their larger size may have allowed them to avoid being entrapped by resin. Although it may be that these two tribes were not present in Europe during the Eocene to early Miocene, there is no phylogenetic evidence that the Sphaeromiini and Stenoxenini are actually Gondwanian groups (Hennig, 1966).

What might explain the striking amount of cladogenesis within the Ceratopogonidae during the very latest Cretaceous and early Tertiary? There are at least several hypotheses to consider. One is the possibility of competitive exclusion, as the mammals likely experienced at the hands (feet?) of the dinosaurs; however, it is difficult to suggest what other group of organisms might have been competitively excluding the Ceratopogonidae prior to the Tertiary. Second, it is significant to note that most of the generic diversification within the Ceratopogonidae is present among taxa in which the female adults feed on other nematocerous Diptera (all those defined by synapomorphy 28'; also likely present in *Paradasyhelea* and *Washingtonhelea*; Fig. 21). It is possible, therefore, that Ceratopogonidae diversified in relation to the increased diversification or abundance of prey type. However, both the Chironomidae and Chaoboridae were diverse and abundant in both the Jurassic and Cretaceous (Jell and Duncan, 1986; Kalugina, 1985, 1986), making a correlated increase in diversification an unlikely hypothesis. Nevertheless, the evolution of sucking blood from insect prey during the Upper Cretaceous may have opened up a wide variety of evolutionary possibilities for Ceratopogonidae attacking these abundant types of prey. The evolution of blood-sucking from insect hosts may have allowed the Ceratopogonidae to utilize a far more accessible and abundant food source (compared with locating and feeding on vertebrates), permitting them to invade a wide array of semiaquatic and aquatic habitats (and thereby diversify further).

The current analysis may also be used to examine the pattern of extinction in the Ceratopogonidae. Although this study characterizes a number of genera as "extinct", this term is somewhat misleading. Of the Cretaceous genera which were recognized solely as fossils (i.e. those labelled as "extinct", see figs 25, 26), none had identifiable synapomorphies indicating their monophyly. As such, logically these may either form sister group relationships with extant lineages, as indicated, or may actually represent ancestral groups and hence lie directly on the phylogenetic line leading to extant taxa. Therefore, there is no evidence of any actual extinction of any lineage of Ceratopogonidae in the Upper Cretaceous.

Within Baltic and Saxonian ambers, the genus *Eohelea* is clearly monophyletic (based on the unique wing patch modifications) and represents the only recognizable case of true extinction within the family at the generic level. Other fossil genera in Baltic and Saxonian amber are questionably monophyletic and hence, may possibly be ancestral to extant taxa.

The lack of evidence for Upper Cretaceous extinction of Ceratopogonidae, so well documented for some vertebrates and ammonites, may not be surprising. First, available evidence indicates that the vertebrate freshwater community in the Upper Cretaceous exhibited little extinction (Sheehan and Fastovsky, 1992; Van Valen, 1984). Ceratopogonidae, with their aquatic or semiaquatic immatures, may have been similarly spared. Second, there is very little evidence (but see below) that any insect group experienced either an extinction event or a marked reduction in diversity at the Cretaceous-Tertiary boundary (Briggs, 1991; Briggs, et al., 1988; Jarzembowski and Ross, 1994; Labandeira and Sepkoski, 1993; Whalley, 1987). Furthermore, some authors analyzing relative rates of generic and familial extinction in insects (e.g. Pike, 1994; Poinar, 1992; Rohdendorf and Zherichin, 1974), fail to distinguish between fossil taxa which are recognizable as truly extinct (i.e. those which have autapomorphies no longer present) and those that are extinct by virtue of not being recognizable in the extant fauna. The latter may merely represent fossil taxa which are ancestral and which have been modified through time (see, for example, Archibald (1992) who showed that some Paleocene mammals previously considered to be extinct had actually been modified and therefore not recognized in younger strata). Of course some taxa may lack identifiable autapomorphies and yet be truly extinct as well but these could only be recognized through an examination on an extremely detailed and chronologically continuous fossil record.

Poinar (1992) has pointed out that there are few examples of transitional groups in the fossil record of modern families of insects. If the punctuated equilibrium model of evolution is valid (rapid morphological changes associated with cladogenesis; Eldredge and Gould, 1972), many of the apparently “extinct” taxa may merely represent ancestral groups. It is unfortunate that so few authors analyzing Cretaceous fossils have put their information within a cladistic framework which would allow for the testing of this idea.

Nevertheless, Heie and Pike (1992) provided limited evidence that at least some groups of Upper Cretaceous aphids are truly extinct at the generic and familial level (some exhibit autapomorphies not present in recent taxa). These authors interpreted such Upper Cretaceous groups as having followed their gymnosperm hosts into extinction. No other studies of Cretaceous amber insects have provided cladistic hypotheses which might indicate whether the "extinct" taxa possessed autapomorphies or whether they might be ancestral. Once these become available it would be additionally interesting to study whether there are differential rates of extinction between terrestrial and freshwater groups of insects as has been reported for vertebrates (Sheehan and Fastovsky, 1992).

The above pattern is consistent with that found by Sheehan and Fastovsky (1992) who concluded that 88% of terrestrial vertebrates but only 10% of freshwater vertebrates went extinct at the end of the Cretaceous. They suggested that the dependence on detritus by freshwater organisms was the key factor which allowed these taxa to survive a major asteroid impact, while those land organisms which depended primarily or secondarily on plants, which themselves experienced a major extinction (Wolfe and Upchurch, 1986), were decimated. Ceratopogonidae larvae are dependent on detritus, small prey such as single-celled organisms, nematodes and the larvae of other insects found in aquatic/semiaquatic habitats; as adults they feed, depending on the species, either on the blood of vertebrates or invertebrates (generally other aquatic midges). As such, ceratopogonids are closely connected trophically to detritus or are directly dependent on organisms feeding on detritus in aquatic/semiaquatic systems. The aphids, the only insect group for which evidence exists of Late Cretaceous extinctions, were closely tied to their plant hosts and hence were more vulnerable.

Evolution of blood-sucking and host selection in Ceratopogonidae

Female adult Ceratopogonidae have a fascinating variety of feeding modes, including taking nectar from flowers, piercing and ingesting the contents of pollen grains, sucking blood from a wide array of other arthropods, or biting vertebrates including fish, reptiles, birds and mammals. A few female ceratopogonids do not feed at all. There are morphological features of the mouthparts which correspondingly vary within the family. To interpret the feeding mode of fossil Ceratopogonidae, the following section analyzes the correlation between feeding mode and mouthpart structure of extant Ceratopogonidae in the light of both outgroup comparisons and cladistic relationships within the family.

Several authors have compared the mouthparts of various Ceratopogonidae within a context of their different feeding modes

(Downes, 1955, 1971, 1978; Gad, 1951; Glukhova, 1981; Goetghebuer, 1923; McKeever, et al., 1991), recognizing morphological correlations with feeding on vertebrate or invertebrate hosts. Cranston (1980) and Downes (1971, 1978) have argued that the mouthparts of biting Nematocerous flies are so similar and the feature so widespread within the suborder that it must be homologous and plesiotypic within the Diptera. Actual study of the character state distributions below indicate a pattern somewhat more complex and difficult to explain.

Table 15 lists the variety of feeding modes and character states of mandibles and lacinia of the families of Nematocera and taxa closely related to the Diptera. It is striking that the habit of sucking blood from either arthropods or vertebrates is present in only eight of 26 families of Nematocera (Ceratopogonidae is not listed in Table 15). The families in Table 15 are arranged in a phyletic sequence according to the cladistic relationships of Nematocera proposed by Wood and Borkent (1989).

The presence of blood-sucking in Siphonaptera is difficult to interpret; they appear to use a fundamentally different mechanism to piece skin membrane (Wenk, 1953). They have lost the mandibles and the laciniae bear teeth and are used as cutting structures. This may indicate that they have acquired the biting habit independently of Nematocerous Diptera.

Nearly all biting Nematocera belong to a monophyletic group composed of the Psychodomorpha, Ptychopteromorpha and Culicomorpha (Tables 15, 16); otherwise only the Blephariceridae are known to suck the blood of other insects. Conceivably, the Blephariceridae acquired their blood-sucking habit independently of other biting Nematocera.

There is one further lineage originating from within the Nematocera which includes biting members. The Orthorraphous Brachycera likely arose from within the Psychodomorpha (Woodley, 1989; Sinclair, 1992b) and early lineages within this group include vertebrate biters with modified mandibles and laciniae (Sinclair et al., 1994).

Considering the widespread habit of blood-sucking from vertebrates in all major lineages within the Culicomorpha (Table 16) (Wood and Borkent, 1989), it seems likely that this is a plesiotypic feature at least within the infraorder. There seem to be three reasonable interpretations of this pattern. First, if the biting habit of Siphonaptera is homologous with that of biting Nematocera, then the biting habit is likely plesiotypic within the Diptera as proposed by Cranston (1980) and Downes (1971, 1978) and then has been lost numerous times within the Diptera; second, if plesiotypic to the Diptera, the feature may have been lost in basal lineages and then reappeared as atavisms in distantly related taxa; third, if the state in Siphonaptera is not homologous it would seem likely that biting mouthparts have evolved twice within the Diptera:

once in Blephariceridae (biting other insects) and once in the ancestor of the Culicomorpha (biting vertebrates).

Crosskey (1990:65) suggested that the earliest Simuliidae had biting mouthparts but that they used these to take sap from plants. The presence of vertebrate blood-sucking in other families of Culicomorpha makes this hypothesis highly unlikely (Table 16).

It seems likely therefore, that the blood-sucking habit is certainly a plesiotypic feature of the Ceratopogonidae and that the biting habit is homologous with that of all other biting Nematocera with the possible exception of the Blephariceridae.

The blood-sucking habit of Ceratopogonidae was discussed by Downes (1971) and Grogan and Szadziewski (1988) within a phylogenetic context. They correctly point out that this habit is likely plesiotypic within Ceratopogonidae and is retained in most early lineages in the family. A summary of the biting habits of Ceratopogonidae is presented in Table 16. Taxa within Ceratopogonidae are arranged in a phyletic sequence (as based on Fig. 21) and support the interpretation of blood-sucking as plesiotypic within the family. Furthermore, because feeding on vertebrates is present in both *Leptoconops* and *Austroconops*, it appears that this condition is plesiotypic within the family with feeding on other arthropods being secondarily derived. The deduced presence of vertebrate blood-sucking in the extinct genus *Atriculicoides* (Figs 22D,25), as the Upper Cretaceous sister group of the Forcipomyiinae + *Dasyhelea*, supports this interpretation.

A number of lineages have lost the biting habit altogether (which is true in every family of biting Nematocera).

Within Nematocera there is a general correlation between the possession of serrated mandibles and the blood-sucking habit (Tables 15, 16). Furthermore, finely toothed mandibles are present in those taxa that feed on vertebrate blood and coarsely toothed mandibles in those that feed on invertebrates (Tables 15, 16). There are several exceptions to this pattern: within the Culicidae, mandibular teeth are present only in the Anophelinae, a relatively early lineage within the family. Their absence in the remaining taxa is likely a loss correlated with the highly elongated and modified mouthparts present in this family. Within the Ceratopogonidae (Table 16), most taxa fit the general pattern (McKeever, et al., 1991); only some *Forcipomyia* have finely toothed mandibles but feed on invertebrates. However, in these taxa the lacinia are strikingly modified with either very small bumps (never as retrorse teeth), ridges or prongs (Wirth, 1991). Many species in *Forcipomyia* and *Atrichopogon* have finely toothed mandibles but have no recorded host.

One Tanyderidae (*Radinoderus occidentalis* Alexander) has finely toothed mandibles (Downes and Colless, 1967) but its feeding mode is not known; based on other modifications, it may feed on frogs (see below).

There is also a correlation between modifications of the lacinia and feeding mode. Nearly all taxa with lacinia bearing retrorse teeth suck blood from vertebrates (Tables 15, 16). However, there are some exceptions to this pattern. First, in some Psychodidae (*Sycorax silacea* Curtis, *Sergentomyia squamipleuris* (Newstead)), in Corethrellidae (McKeever, 1986; McKeever and Pound, 1979), and in one lineage of *Forcipomyia* (Glukhova, 1981) which suck vertebrate blood, females have finely toothed mandibles but lack toothed laciniae. These are also the only groups of Nematocera, however, which feed on frogs (Desportes, 1942; Lewis and Lane, 1976; McKeever and Pound, 1979). The only exception to this characterization of frog feeders are some species of Culicidae (e.g. some *Aedes* Meigen, some *Culex* Linnaeus), which have smooth mandibles and toothed maxillae (these generally also feed on other vertebrates); this is likely best explained by the otherwise highly modified mouthparts of the Culicidae. A further exception to this pattern is found within the Psychodidae; in *Phlebotomus* (*Idiophlebotomus* Quate and Fairchild) females have serrated mandibles and smooth or nearly smooth maxillae. They feed on bats but perhaps secondarily on frogs (Duckhouse, pers. comm.).

Within three lineages of Ceratopogonidae, some *Atrichopogon*, some *Forcipomyia* and a few species of *Culicoides* have laciniae with retrorse teeth but feed on invertebrates. However, in nearly all these species, the mandibles are coarsely toothed, a feature otherwise correlated with blood-sucking from invertebrates. Only a few species of *Culicoides* (*Trithecoides* Wirth and Hubert) are known (Wirth and Hubert, 1989) to have mandibles with coarse teeth and to feed on both vertebrates and invertebrates.

To summarize, the above observations indicate that the combination of finely toothed mandibles and laciniae with retrorse teeth indicates that the female adult sucks blood from vertebrates. Furthermore, the presence of coarse mandibular teeth indicates that the female likely sucks blood from an invertebrate host.

Within the Nematocera, only three families include species which have strongly toothed labra (Table 16). In some Psychodidae the apex of the labrum is armed with well developed teeth (Lewis, 1975). In the Simuliidae, the female labrum is distinctive, with several hooked teeth located at the its very apex (Sutcliffe and McIver, 1984). The labra found in both of these families are quite different from the toothed labrum found in a few *Forcipomyia* and all biting *Culicoides*, where the margin of the apical portion of the labrum has a continuous row of teeth (Jobling, 1928; McKeever et al., 1988; Sutcliffe and Deepan, 1988). The presence of a toothed labrum therefore is somewhat homoplastic but may be used in combination with other characters to identify these taxa. Some other Ceratopogonidae have small teeth or spicules on the labrum (McKeever, et al., 1991), which in whole mounts look like small bumps or hair-like spicules.

Finally, only a few Nematocera have a toothed hypopharynx. These are present in some Psychodidae (Lewis, 1975), in Corethrellidae (McKeever, 1986; McKeever and Pound, 1979), possibly some Simuliidae (where the teeth appear apically spiculate) (Sutcliffe and McIver, 1984), and within Ceratopogonidae in *Austroconops*, some *Atrichopogon* and in all biting *Culicoides* (McKeever, et al., 1988). A toothed hypopharynx shows some homoplasy but may be used in combination with other mouthpart character states to identify these taxa (Table 16).

The presence of a toothed labrum and hypopharynx is correlated with the habit of sucking blood from vertebrates. However, some vertebrate blood feeders (i.e. *Leptoconops, Austroconops*) do not have both features.

It is unfortunate that the feeding habits of a few *Forcipomyia* with a toothed labrum and some *Atrichopogon* with a toothed hypopharynx are unknown.

These observations allow for an interpretation of the feeding habits of Canadian Cretaceous fossils (Table 17). The presence of both finely toothed mandibles and laciniae with teeth indicates that several species fed on vertebrate blood. The mandible of the unnamed species of *Atriculicoides* could not be seen but the presence of a toothed labrum and lacinia indicates that it too, fed on vertebrate blood.

Two species had coarsely toothed mandibles and therefore probably sucked blood from invertebrate hosts: *Peronehelea chrimikalydia* and *Palaeobrachypogon aquilonius.*

The details of the mouthparts of female *Palaeobrachypogon remmi* and *P. vetus* were not visible and could not, therefore, be interpreted. However, the large tarsal claws of both these species indicates that they likely sucked blood from other insects. This feature is restricted in extant species to those which prey on insects. However, within that group, some members have lost the biting habit and still retain large claws. It is possible, therefore, that *P. remmi* and *P. vetus* either preyed on other insects or were non-biting.

Two of the three genera identified as sucking blood from vertebrates are living today.

Extant species of *Leptoconops* are known to feed on a variety of vertebrate hosts including reptiles, birds and mammals (Auezova et al., 1990; Mayer, 1955); their mouthparts do not appear to have diversified to reflect this variation in host type. Details of the mouthparts of two female *L. primaevus* were visible. In both specimens the mandibles were finely serrated, with small teeth and the laciniae had at least six retrorse teeth. These features indicate that the females of *L. primaevus* sucked blood from some unidentified vertebrate in southern Alberta during the Late Cretaceous.

Extant *Culicoides* also feed on a wide array of hosts, encompassing mosquitoes, fish, reptiles, birds and mammals (Blanton and Wirth, 1979; Wirth and Hubert, 1989). At least two features of their morphology appear to vary with the type of host.

The pattern of sensilla coeloconica on the antennae varies within extant *Culicoides* in relation to whether they feed on birds or mammals (Blackwell, et al., 1992b; Braverman and Hulley, 1979; Chu-Wang et al., 1975; Felippe-Bauer, et al., 1989; Jamnback, 1965). Ornithophilic species have sensilla coeloconica on 8-13 of their flagellomeres and mammalophilic species have them on only 4-6 flagellomeres. It appears that Canadian Cretaceous *Culicoides* have sensilla coeloconica only on the first flagellomere and therefore this characteristic cannot be used to interpret the host type of the fossil species. Because Wirth and Navai (1978) hypothesized that these sensilla may function as heat receptors, required for host detection, it is possible that the lack of sensilla coeloconica on flagellomeres 2-13 indicates that their hosts were cold-blooded (such sensilla are known heat receptors in mosquitoes (Davis and Sokolove, 1975)). Considering the conclusion below that at least some species of Cretaceous *Culicoides* fed on large dinosaurs, this may further indicate that those hosts were cold-blooded as well. However, it is more likely these species of *Culicoides* belong to a lineage in which the character state of more extensively distributed sensilla coeloconica had not yet evolved (see discussion of *Culicoides* under "Phylogenetic Relationships of Cretaceous Ceratopogonidae"). Certainly all other genera of Ceratopogonidae which feed on warm-blooded vertebrates do not have sensilla coeloconica at all or have any beyond the first flagellomere (i.e. *Leptoconops*, *Austroconops*, some *Forcipomyia*; note that these genera have at least one type of sensilla on the first flagellomere: trichodea, basiconica and ampullacea, and that these are of unknown function) (Borkent, et al., 1987:600). In addition, those extant species of *Culicoides* which feed on mosquitoes or fish (Wirth and Hubert, 1989) have 4-10 antennal flagellomeres with sensilla coeloconica (Wirth and Hubert, 1989), indicating that merely feeding on cold-blooded hosts does not eliminate the more extensively distributed sensilla. Finally, Blackwell et al. (1992b) have indicated that sensillae coeloconica generally respond to $C0_2$, temperature and/or humidity and it is clear that further study is required to determine specific function in extant species.

The second feature which appears to vary with host type in extant species of *Culicoides* is the third maxillary palpal segment which bears a patch of capitate sensilla. These sensilla vary in number as well as in location: from being situated on the unmodified surface of the palpal segment to being located in a deep pit (Figs 27A-AX). Although the function of the capitate sensilla has not been determined in ceratopogonids, their morphology is very similar to those in related families (Chu-Wang, et al., 1975; Sutcliffe et al., 1987) and they are almost certainly homologous in these groups. The sensilla are known to be receptors for CO_2 in Culicidae (Kellogg, 1970). The sensilla are likely important in locating the host from a distance; the fly tracks the host by following the CO_2 plume which flows downwind from the host (Sutcliffe, 1987).

Rowley and Cornford (1972) suggested that the number of the capitate sensilla in species of *Culicoides* is related to host size. In their study, two species which fed on small hosts (i.e. birds and small mammals) had a greater number of sensilla (n= 29-74) than did those two species which fed on large mammals (sensilla number = 11-17). One of the species they studied, *Culicoides guttipennis* (Coquillett), feeds on both birds and large mammals (i.e. man) and had a range of 29-36. Three other species which feed on large mammals have a low number of capitate sensilla: *C. furens* (Poey) has 9-12 (Chu-Wang et al., 1975), *C. nubeculosus* (Meigen) has 12-17 (Messaddeq et al., 1989) and *C. hollensis* (Melander and Brues) has 11 (Linley, 1976:343). Braverman and Hulley (1979) showed a similar pattern for seven African species; species which fed on large mammals (or predominately large mammals) had 9-24 sensilla while ornithophilic species had 33-75 sensilla. In summary, species of *Culicoides* which feed on large mammals have 9-24 capitate sensilla while those which feed on birds have 29-75. This variation in number likely reflects the need to detect lesser (needing more capitate sensilla) or greater (needing fewer capitate sensilla) concentrations of CO_2 emanating from smaller or larger hosts, respectively. Unfortunately, these observations are based on a total of only 15 species. It would be invaluable for further studies to be made in this regard.

A similar correlation between number of capitate sensilla and host size has been discussed for a few species of Simuliidae (Mercer and McIver, 1973). McIver and Charlton (1970) reported variation in the number of capitate sensilla for a few species of mosquitoes; examination of host size preferences (Wood et al., 1979) indicates a similar correlation within the Culicidae. Although these reports were based on only a few species each, the fact that the sensilla are homologous with those of Ceratopogonidae further supports the hypothesis that the number of sensilla are related to CO_2 concentrations emanating from hosts of varying size.

The few species of *Culicoides* which feed on mosquitoes also have a low number of capitate sensilla (<19) but these species are distinguished by the presence of coarse teeth on the mandible, reflecting their habit of sucking blood from other insects (Glukhova, 1982; Wirth and Hubert, 1989). *Culicoides cylindripalpis* also likely feeds on other insects, based on its coarsely toothed mandible, but has 39 capitate sensilla (Wirth and Hubert, 1989). However, these sensilla are unusual in being scattered over the entire surface of the third palpal segment.

At least two species of Canadian Cretaceous *Culicoides* have a low number of capitate sensilla: *C. canadensis* (n= 10) and *C. bullus* (n= about 7). These observations indicate that these two species fed on large vertebrates.

Further evidence on host type is apparent in the shape of the third maxillary palpal segment itself. Figs 27A-AX illustrates the maxillary palps of many Nearctic species of *Culicoides*. The sketches were first arranged according to host size and type and then sorted from most slender to most expanded. Figs 27A-Q are those which feed on large mammals (e.g. horses, cattle, man, hogs), Figs 27R-AE are those which feed on both large mammals and smaller vertebrates (including small mammals and/or birds), Figs 27AF-AM are those which feed only on birds, and Fig. 27AN is of the only species which feeds on turtles. Finally, Figs 27AO-AX are of those species which do not feed at all.

It is possible that this feature is merely a correlate with the number of capitate sensilla discussed above (i.e. species with fewer capitate sensilla have a more slender third maxillary palpal segment). However, this may not be the case. For example, *C. variipennis* (Coquillett) may have a relatively thick palpal segment (Figs 27P) but has only 11-17 sensilla; similarly, *C. guttipennis* has a relatively slender palpal segment (Fig. 27R) and has 29-36 sensilla (Rowley and Cornford, 1972).

The scoring of host type was more difficult than is apparent from Figs 27A-AX, reflecting some flexibility in some species. In general, I have appraised each species according to the number of host records known. If there was some question of whether a species fed on both big and small mammals and/or birds, I placed that species in that category. If I considered a given record as rare and exceptional, then the species was placed in the category indicated by the bulk of the records. For example, *C. biguttatus* (Coquillett) and *C. furens* feed primarily on big mammals but are also known from a very few records to feed on birds. I have considered these records to be exceptional and have scored them as big mammal feeders. Similarly, one species, *C. stellifer* (Coquillett), is known to feed commonly on birds but is also known from a number of records as feeding on large mammals; I placed that species in the mixed category.

On the basis of these illustrations, two patterns emerge. First, most species which do not bite have relatively short and squat third palpal segments. Secondly, elongate and slender third palpal segments are characteristic of those species that feed on big mammals. There are three discrepancies in this pattern and these are present amongst species which appear to feed on both big mammals and birds: *C. guttipennis* (Fig. 27R) appears to be a general feeder and is recorded from big mammals, rabbits, turkeys and chickens; *C. sanguisuga* (Coquillett) (Fig. 27S) feeds on a variety of hosts but appears to prefer big mammals; *C. arboricola* Root and Hoffman (Fig. 27T) appears to feed primarily on birds, with only one record of feeding on man and another on rabbits.

These observations imply that an elongate, slender palpal segment is a likely indicator of species which feed on big mammals; out of nine species with slender palpal segments, eight are known to be entirely or primarily big mammal feeders.

This pattern suggests that the elongate, slender palpal segment of *C. filipalpis* was related to feeding on a large vertebrate. However, the palpal segment of *C. filipalpis* is proportionally more elongate that any known extant species of *Culicoides*. At best, I consider the use of the palpal segment shape as only weak evidence for determining host type and a hypothesis which requires further testing.

I conclude therefore, that *C. canadensis, C. bullus* and possibly *C. filipalpis* fed on large vertebrates (i.e. a host which produced large amounts of CO_2 as evidenced by the low number of capitate sensilla on the third maxillary palpal segment). During the Late Cretaceous, mammals were relatively small (not larger than the North American opossum) (R.C.Fox, pers. comm.) and although birds were well diversified by the Late Cretaceous (Sereno and Chenggang, 1992), none were larger than modern pelicans and loons (Romer, 1966; Russell, 1989). As such, the only likely candidates as hosts for these species of fossil *Culicoides* were the dinosaurs which were abundant, both in species and numbers, in southern Alberta during this time (see "Palaeoecology" for further discussion). Of the more than 30 types of dinosaurs, at least 19 weighed more than 150 kg and the most abundant of these were the hadrosaurs and ceratopsians.

It was not possible to determine the number of sensilla on the maxillary palpus of *C. tyrrelli* and thereby deduce its host using this character state. The shape of its third maxillary palpal segment (Fig. 9L) is characteristic of extant species which feed on both large and small vertebrates (Figs 27A-AN).

If *C. canadensis, C. bullus* and *C. filipalpis* were feeding on dinosaurs, where might have been their specific site of attack on the dinosaur's body? There are very few observations of ceratopogonids feeding on reptiles, which are probably the closest models we have of the challenges presented to biting ceratopogonids acquiring a blood-meal from an organism with scales. Auezova et al. (1990) recorded *Leptoconops* (the earliest lineage of Ceratopogonidae, Fig. 21) feeding on the sides of the stomach, neck, head and extremities of lizards in Kazakhstan. They noted that lizards shedding their skins were particularly vulnerable (probably due to the relatively soft nature of the new surface of their skin and scales). They did not describe how the female midges circumvented the difficulty presented by the overlapping scales which cover most of the lizard's bodies.

Female *C. testudinalis* Wirth and Hubert feed on turtles (Wirth and Hubert, 1962) but there are no published observations of further details. Females of a species of *Forcipomyia* (*Lasiohelea*) feed on iguanas in Costa Rica (pers. obs.; apparently on the head but details not observed) and female *C. phlebotomus* (Williston) feed on the heavily vascularized soft shell of female leatherback turtles as these are on the beach to lay their eggs (pers. obs. in Costa Rica; ms. in prep.).

The second oldest lineage of Ceratopogonidae (Fig. 21) is represented by one extant species of *Austroconops* found in southwestern Australia. Females were recorded as feeding on human eyelids (Wirth and Lee, 1958).

There are probably at least two basic requirements regarding the specific site on a host from which Ceratopogonidae may suck blood: access to the skin surface (i.e. no scales, feathers or fur) and heavy vascularization in that area. It is probably that past selection combined with a degree of host specificity has led to species specializing on certain specific feeding sites on a given host. For example, Meiswinkel (pers. comm.) notes that at least five species of *Culicoides* feed behind the ears of African elephants, an heavily vascularized area with soft, thin skin.

The skin of dinosaurs has been preserved for only a few species and appears to be composed of non-overlapping scutes, separated by narrow fields of tiny scutes (D.A. Russell, pers. comm). The membrane between these skutes may have been a site of attack for biting ceratopogonids. However, it is virtually certain that the eyelids of dinosaurs included a significant amount of inter-scale membrane with a rich blood supply. These may also have been a feeding site for females of extinct *Culicoides*, much as extant female *Austroconops mcmillani* do today.

There is no evidence that dinosaurs shed their skin during growth (Russell, pers. comm.) making them more susceptible to biting midges as is the case with modern lizards.

These inferences are summarized in the conjectural frontispiece to this article, showing female *C. canadensis* feeding on the eyelids of an adult duck-billed dinosaur of the genus *Corythosaurus*. These dinosaurs frequented the margins of aquatic habits (Russell, 1989) where at least some *Culicoides* likely bred. Russell (pers. comm.) has also pointed out that the underside of ceratopsian neck frills were vascularized and may have also provided feeding sites for these biting midges.

A few previous authors have conjectured that ceratopogonids have a long history of feeding on vertebrates (Downes, 1971, 1978:58; McAlpine and Martin, 1969) and even on dinosaurs (Grogan and Szadziewski, 1988; Pellegrino, 1985). It is interesting that data are now available to support the suggestion of blood-sucking of dinosaurs.

Glukhova (1989:101) illustrated the mouthparts of Late Cretaceous *C. kaluginae* from Taimyr amber, showing finely serrated mandibles and laciniae with retrorse teeth. This indicates that this species fed on vertebrate blood but, because the capitate sensilla are not reported (probably because they could not be seen) and the third maxillary palpal segment is swollen, it is not possible to deduce a more specific host.

It is also not possible to identify the specific hosts of species of *Atriculicoides*. Although the presence of finely toothed mandibles and laciniae with retrorse teeth indicates that these were vertebrate feeders, the fact that the genus is extinct makes it impossible to presently interpret further details of mouthpart structure which might be correlated with a more specific host type. Szadziewski and Schlüter (1992) have previously hypothesized, based on unmodified legs with short claws, rudimentary empodia and the presence of numerous sensilla on the third palpal segment that the Cretaceous *Atriculicoides* species they studied from France sucked blood from vertebrates. My observations of their mouthparts here confirms that conjecture.

The phylogenetic conclusions and interpretation of Upper Cretaceous Ceratopogonidae above ("Patterns of diversification, species turnover and extinction in Ceratopogonidae and other insects") indicate that feeding on the blood of other invertebrates is a derived condition and that this feeding strategy has evolved at least three times within the family.

Members of the Forcipomyiinae feed only on insects or arachnids which are significantly larger than themselves (e.g. Orthoptera, Coleoptera, caterpillars, the wing veins of some Odonata and Neuroptera); they essentially attack them in a similar manner as the vertebrate feeders which attack their much larger hosts. That this feeding mode has evolved independently from other lineages feeding on nematocerous Diptera is further supported by the discovery that *Atriculicoides*, as the Cretaceous sister group of the Forcipomyiinae + Dasyheleinae, likely fed on vertebrate blood (see below under "Evolution of blood-sucking and host selection in Ceratopogonidae").

A few species of *Culicoides* (*Trithecoides*) have been recorded sucking blood from the abdomens of mosquitoes (Wirth and Hubert, 1989), attacking these in a similar mode as do the Forcipomyiinae (i.e. as smaller ceratopogonids on a large insect host) rather than as most Ceratopogoninae (which capture and feed on other Nematocerous Diptera of equal or smaller size). The widespread occurrence of vertebrate blood feeding in other *Culicoides* suggests that the feeding behaviour of some species of *Culicoides* (*Trithecoides*) has been acquired independently.

Finally, members of the lineage defined by synapomorphy 28' suck blood from other Nematocerous Diptera (generally Chironomidae or Chaoboridae) of equal or smaller size, capturing these in flight (Downes, 1978). Female *Paradasyhelea* and *Washingtonhelea*, based on their mouthparts, likely also feed on other insects, indicating that the evolution of feeding on other nematocerous Diptera may have actually arisen earlier than the occurrence of synapomorphy 28".

Feeding on other insects has therefore evolved at least three times in the Ceratopogonidae: within the Forcipomyiinae, a few species of *Culicoides* and in those Ceratopogoninae defined by synapomorphy 28' (Fig. 21). Differences in details of mode of attack and type of host confirm their independent evolution.

There is initial evidence that the group defined by synapomorphies 17" and 26 (Fig. 21) originated between 83 and 99 million years ago (see 'Patterns of diversification, species turnover and extinction in Ceratopogonidae and other insects' pp. 124). If true, this pattern would show that the lineage of ceratopogonids which feed as adult females on the blood of other nematocerous Diptera (defined by synapomorphy 28' in Fig. 21) likely arose during the Upper Cretaceous.

Downes (1971) discussed the variation of feeding modes in Forcipomyiinae (some also feed on pollen grains) and suggested that such intrageneric flexibility might indicate a plesiotypic condition characteristic of the ancestor of the Diptera. However, the presence of vertebrate blood-sucking in *Leptoconops, Austroconops*, and *Atriculicoides* indicates that vertebrate blood-sucking is plesiotypic for this group; the variation in host type in the Forcipomyiinae and the lost of blood-sucking in *Dasyhelea* are likely secondary modifications.

SUGGESTIONS FOR FUTURE RESEARCH

This investigation indicates several important areas which require further investigation.

This analysis of the fossil Ceratopogonidae illustrates how vital it is to understand the cladistic relationships between extant members of a group to adequately interpret its fossil members. Much work remains to be done interpreting the characters and distribution of character states in ceratopogonids. The features of larvae and pupae, in particular, are virtually untouched and will provide critical information for interpreting phylogenetic relationships (pers. obs.).

It was striking that so few studies (n= 18) are available interpreting Upper Cretaceous Canadian amber arthropods. There are many specimens of many taxa (McAlpine and Martin, 1969; Pike, 1994) in museums awaiting analysis (especially mites and Chironomidae). Although Poinar, 1992:66 indicated that the CNCI (as the Biosystematics Research Institute) has only 300 pieces of amber with insects, in fact there are at least 1,281 (J.M. Cumming, pers. comm.). Similarly, the Museum of Comparative Zoology has, in addition to its main collection, many kilograms of Canadian amber which await examination and which will certainly turn up many more fossils than the 5,000 reported.

The analysis of the biting habits of fossil Ceratopogonidae also pointed out major gaps in our understanding of the mouthparts of biting flies. Most of our information is based on a handful of species. Certainly, future comparative studies will test some of the ideas presented here.

Finally, the recognition that there is little evidence of extinction of insects at the Cretaceous-Tertiary boundary has important implications for life today. Of all terrestrial and freshwater life, it appears that the major extinction event at the end of the Cretaceous primarily involved larger organisms (i.e dinosaurs, marsupials; Briggs, 1991); much of the remaining ecosystem likely was largely unharmed. Our abuse of life on Earth today is resulting in the destruction of a high number of insects and other smaller life forms and this extinction event may be without precedence in the history of our planet. We are attacking the base of the food web and are certainly unravelling an exceedingly complex ecosystem made up primarily of relatively small organisms. This frightening process bodes poorly for our own existence and that of many other forms of life on our planet.

REFERENCES

Archibald, J.D. 1992. The importance of phylogenetic analysis for the assessment of species turnover: a case history of Paleocene mammals in North America. Paleobiology 19:1-27.

Arnett, R.H. and G.A. Samuelson 1986. The insect and spider collections of the world. E.J. Brill/Flora and Fauna Pub., Gainesville, Florida, 220 pp.

Ashe, P. 1983. A catalogue of chironomid genera and subgenera of the world including synonyms (Diptera: Chironomidae). Entomologica Scandinavica Supplement 20:1-68.

Atchley, W.R. 1967. The *Culicoides* of New Mexico (Diptera: Ceratopogonidae). University of Kansas Science Bulletin 46: 937-1022.

Atchley, W.R. 1970. A biosystematic study of the subgenus *Selfia* of *Culicoides* (Diptera: Ceratopogonidae). University of Kansas Science Bulletin 49:181-336.

Atchley, W.R. and R.C. Jackson. 1968. Techniques and preliminary studies on the chromosomes of *Nilobezzia* and *Bezzia*. Annals of the Entomological Society of America 61: 1524-1527.

Auezova, G., Z. Brushko, and R. Kubykin. 1990. Feeding of biting midges (Leptoconopidae) on reptiles. Second International Congress of Dipterology. Abstract Volume, pp. 12.

Bakker, R.T. 1972. Anatomical and ecological evidence of endothermy in dinosaurs. Nature 238:81-85.

Baroni-Urbani, C. and J.B. Saunders. 1982. The fauna of the Dominican Republic amber: the present status of knowledge. Proceedings of the Ninth Caribbean Conference, Santo Domingo, pp. 213-223.

Barr, W.B. 1984. Prolegs and attachment of *Simulium vittatum* (sibling IS-7) (Diptera: Simuliidae) larvae. Canadian Journal of Zoology 62:1355-1362.

Barron, E.J. 1987. Global Cretaceous paleogeography - International Geologic Correlation Program Project 191. Palaeogeography, Palaeoclimatology, Palaeoecology 59:207-214.

Beardsley, J.W. 1969. A new fossil scale insect (Homoptera: Coccoidea) from Canadian amber. Psyche 76:270-279.

Béland, P. and D.A. Russell. 1979. Ectothermy in dinosaurs: paleoecological evidence from Dinosaur Provincial Park, Alberta. Canadian Journal of Earth Sciences 16:250-255.

Bell, W.A. 1965. Upper Cretaceous and Paleocene plants of western Canada. Geological Survey of Canada Paper 65-35, 46 pp.

Bidlingmayer, W.L. 1961. Field activity studies of adult *Culicoides furens*. Annals of the Entomological Society of America 54:149-156.

Blackwell, A., A.J. Mordue (Luntz), M.R. Young and W. Mordue. 1992a. Bivoltinism, survival rates and reproductive characteristics of the Scottish biting midge, *Culicoides impunctatus* (Diptera: Cerato-pogonidae) in Scotland. Bulletin of Entomological Research 82:299-306.

Blackwell, A., A.J. Mordue (Luntz), and W. Mordue. 1992b. Morphology of the antennae of two species of biting midge: *Culicoides impuncatatus* (Goetghebuer) and *Culicoides nubeculosus* (Meigen) (Diptera, Ceratopogonidae). Journal of Morphology 213:85-103.

Blanton, F.S. and W.W. Wirth. 1979. The sand flies (*Culicoides*) of Florida. Arthropods of Florida and Neighboring Land Areas 10, xv + 204 pp.

Boesel, M.W. 1937. Order Diptera. Family Chironomidae. pp. 44-55, *In* Insects and arachnids from Canadian amber, by F.M. Carpenter, J.W. Folsom, E.O. Essig, A.C. Kinsey, C.T. Brues, M.W. Boesel, and H.E. Ewing. U. of Toronto Studies, Geological Series No. 40:7-62.

Borkent, A. 1979. Systematics and bionomics of the species of the subgenus *Schadonophasma* Dyar and Shannon (*Chaoborus*, Chaoboridae), Diptera). Quaestiones Entomologicae 15:122-255.

Borkent, A. 1984. The systematics and phylogeny of the *Stenochironomus* complex (*Xestochironomus, Harrisius*, and *Stenochironomus*) (Diptera: Chironomidae). Memoirs of the Entomological Society of Canada. 128. 269 pp.

Borkent, A. 1991. The Ceratopogonidae (Diptera) of the Galápagos Islands, Ecuador with a discussion of their phylogenetic relationships and zoogeographic origins. Entomologica Scandinavica 22:97-122.

Borkent, A. 1992. A new key to some genera of Ceratopogonini in the Holarctic (Diptera: Ceratopogonidae). Entomologica Scandinavica 22:433-436.

Borkent, A. 1993. A world catalogue of fossil and extant Corethrellidae and Chaoboridae (Diptera), with a listing of references to keys, bionomic information and descriptions of each known life stage. Entomologica Scandinavica 24: 1-24.

Borkent, A. and B. Bissett. 1990. A revision of the Holarctic species of *Serromyia* Meigen (Diptera: Ceratopogonidae). Systematic Entomology 15:153-217.

Borkent, A. and D.A. Craig. 1994. The structure and function of the abdominal eversible sacs of female *Bezzia varicolor* (Coquillett) (Ceratopogonidae: Diptera). Canadian Entomologist 126:533-541.

Borkent, A. and W.L. Grogan (in press). A revision of the genus *Ceratopogon* Meigen with a discussion of phylogenetic relationships, zoogeography and bionomic divergence (Diptera: Ceratopogonidae). Memoirs of the Entomological Society of Washington.

Borkent, A. and S. McKeever. 1990. First instar larvae of *Corethrella appendiculata* Grabham have a prothoracic proleg (Diptera: Corethrellidae). Entomologica Scandinavica 21:219-223.

Borkent, A., W.W. Wirth, and A.L. Dyce. 1987. The newly discovered male of *Austroconops* (Ceratopogonidae: Diptera) with a discussion of the phylogeny of the basal lineages of the Ceratopogonidae. Proceedings of the Entomological Society of Washington 89:587-606.

Boorman, J. and R.P. Lane. 1979. A new genus and species of Ceratopogonidae (Diptera) closely related to *Culicoides* from West Africa. Journal of Natural History 13:327-332.

Botosaneanu, L. and W. Wichard. 1983. Upper-Cretaceous Siberian and Canadian amber caddisflies (Insecta: Trichoptera). Bijdrangen tot de Dierkunde 53:187-217.

Braverman, Y. and P.E. Hulley. 1979. The relationship between the numbers and distribution of some antennal and palpal sense organs and host preference in some *Culicoides* (Diptera: Ceratopogonidae) from southern Africa. Journal of Medical Entomology 15:419-424.

Briggs, D.E.G., R.A. Fortey, and E.N.K. Clarkson. 1988. Extinction and the fossil record of the arthropods. pp. 171-209, *In* G.P. Larwood, ed. Extinction and survival in the fossil record. Clarendon Press, Oxford, England.

Briggs, J.C. 1991. A Cretaceous-Tertiary mass extinction? Bioscience 41:619-624.

Brodie, P.B. 1845. A history of the fossil insects in the secondary rocks of England, accompanied by a particular account of the strata in which they occur, and of the circumstances connected with their preservation. xxviii + 130 pp, 11 pls., London.

Broughton, P.L. 1974. Conceptual frameworks for geographic-botanical affinities of fossil resins. Canadian Journal of Earth Sciences 11:583-594.

Brouwers, E.M., W.A. Clemens, R.A. Spicer, T.A. Ager, L.D. Carter, and W.V. Sliter. 1987. Dinosaurs on the North Slope, Alaska; high latitude, latest Cretaceous environments. Science 237:1608-1610.

Brown, B.V. and E.M. Pike. 1990. Three new fossil phorid flies (Diptera: Phoridae) from Canadian Late Cretaceous amber. Canadian Journal of Earth Sciences 27:845-848.

Brundin, L. 1966. Transantarctic relationships and their significance, as evidenced by chironomid midges, with a monograph of the subfamilies Podonominae and Aphroteniinae and the austral Heptagyiae. Kungl. Svenska Vetenskapsakademiens Handlingar. Fjarde Serien. 11(1), 472 pp, 30 pls.

Cameron, A.E. 1947. Scottish blood-sucking midges. Transactions of the Highland and Agricultural Society of Scotland 59:65-76.

Cameron, A.E. 1948. Scottish blood-sucking midges. Transactions of the Highland and Agricultural Society of Scotland 60:58-79.

Campbell, J.A. and E.C. Pelham-Clinton. 1960. A taxonomic review of the British species *Culicoides* Latreille (Diptera, Ceratopogonidae). Proceedings of the Royal Society of Edinburgh (B) 67:181-302.

Chu-Wang, I., R.C. Axtell and D.L. Kline. 1975. Antennal and palpal sensilla of the Sand Fly *Culicoides furens* (Poey) (Diptera: Ceratopogonidae). International Journal of Insect Morphology and Embryology 4:131-149.

Clastrier, J. 1989. Ceratopogonidae de Nouvelle-Calédonie. VII. Genre *Paradasyhelea* [Diptera, Nematocera]. Revue Francaise d'Entomologie 11:133-137.

Clastrier, J. and W.W. Wirth. 1978. The *Leptoconops kerteszi* complex in North America (Diptera: Ceratopogonidae). United States Department of Agriculture Technical Bulletin No. 1573. 58 pp.

Cockerell, T.D.A. 1919. Insects in Burmese amber. Entomology 52:241-243.

Cockerell, T.D.A. 1921. LVII. - Fossil arthropods in the British Museum. - VI. Oligocene insects from Gurnet Bay, Isle of Wight. Annals and Magazine of Natural History Series 9, 7:453-480.

Cook, E.F. 1956. The Nearctic Chaoborinae (Diptera: Culicidae). University of Minnesota Agricultural Experiment Station Technical Bulletin 218, 102 pp.

Cooper, K.W. 1964. The first fossil Tardigrade: *Beorn leggi* Cooper, from Cretaceous amber. Psyche 71:41-48.

Coquillett, D.W. 1902. New Diptera from North America. Proceedings of the United States National Museum 25:83-126.

Coquillett, D.W. 1910. The type-species of the North American genera of Diptera. Proceedings of the United States National Museum 37:499-647.

Craig, D.A. 1977. A re-assessment of the systematic position of *Pseudosimulium humidum* (Westwood), an Upper Jurassic fossil dipteran. Entomologist's Gazette 28:175-179.

Cranston, P.S. 1980. Insects of Saudi Arabia. Diptera: Fam. Chaoboridae. Fauna of Saudi Arabia 2:286-290.

Cranston, P.S. and C.J. Humphries. 1988. Cladistics and computers: a chironomid conundrum? Cladistics 4:72-92.

Cranston, P.S., D.H.D. Edward and D.H. Colless. 1987. *Archaeochlus* Brundin: a midge out of time (Diptera: Chironomidae). Systematic Entomology 12:313-334.

Crosskey, R.W. 1987. 32. An annotated checklist of the world Black Flies (Diptera: Simuliidae). pp. 425-520 *In* Black Flies. Ecology, Population Management, and Annotated World List, eds. K.C. Kim and R.W. Merritt, xv + 528 pp., Pennsylvania State University.

Crosskey, R.W. 1990. The natural history of Blackflies. Wiley & sons, Chichester, New York, Brisbane, Toronto, Singapore. ix + 711 pp.

Davies, K.L. 1987. Duck-billed dinosaurs (Hadrosauridae, Ornithischia) from the North Slope of Alaska. Journal of Paleontology 61:198-200.

Davis, D.R. 1989. An exceptional fossil amber collection acquired by the Smithsonian Institution. Proceedings of the Entomological Society of Washington 91:545-550.

Davis, E.E. and P.G. Sokolove. 1975. Temperature responses of antennal receptors of the mosquito, *Aedes aegypti*. Journal of Comparative Physiology 96:223-236.

Debenham, M.L. 1987a. The biting midge genus *Forcipomyia* (Diptera: Ceratopogonidae) in the Australasian Region (exclusive of New Zealand). I. Introduction, key to subgenera, and the *Thyridomyia* and *Trichohelea* groups of subgenera. Invertebrate Taxonomy 1:35-119.

Debenham, M.L. 1987b. The biting midge genus *Forcipomyia* (Diptera: Ceratopogonidae) in the Australasian Region (exclusive of New Zealand). II. *Warmkea* and the *Caloforcipomyia* group of subgenera. Invertebrate Taxonomy 1:167-199.

Debenham, M.L. 1988. *Chimaerohelea*, a new genus of Ceratopogonidae (Diptera) from North Queensland. Invertebrate Taxonomy 1:801-806. (1987)

Debenham, M.L. and W.W. Wirth. 1984. Australian and New Guinea species of the *Forcipomyia* subgenus *Euprojoannisia* (Diptera: Ceratopogonidae). Australian Journal of Zoology 32:851-889.

Desportes, C. 1942. *Forcipomyia velox* Winn. et *Sycorax silacea* Curtis, vecteurs d'*Icosiella neglecta* (Diesing), filaire commune de la grenouille verte. Annales de Parasitologie Humaine et Comparee 19:53-68.

Dixon, D. 1989. The big book of dinosaurs. Bison Books Ltd., London. 128 pp.

Dorf, E. 1942. Upper Cretaceous floras of the Rocky Mountain Region. II. Flora of the Lance Formation and its locality, Niobrara Co., Wyoming. Carnegie Institute Washington Publication 508:79-159.

Downes, J.A. 1955. The food habits and description of *Atrichopogon pollinivorus* sp. n. (Diptera: Ceratopogonidae). Transactions of the Royal Society of London 106:439-453.

Downes, J.A. 1971. The ecology of blood-sucking Diptera: an evolutionary perspective. pp. 232-258 *In* A.M. Fallis (ed.) Ecology and physiology of parasites: a symposium. Toronto University Press, Toronto.

Downes, J.A. 1978. Feeding and mating in the insectivorous Ceratopogoninae (Diptera). Memoirs of the Entomological Society of Canada 104, 62 pp.

Downes, J.A. 1983. Two weeks on Vancouver Island, in search of biting midges (Ceratopogonidae). Boreus, Newsletter of the Entomological Society of British Columbia 3(2):3-6.

Downes, J.A. and D.H. Colless. 1967. Mouthparts of the biting and blood-sucking type in Tanyderidae and Chironomidae (Diptera). Nature 214:1355-1356.

Downes, J.A. and W.W. Wirth. 1981. 28. Ceratopogonidae. pp. 393-421 in Manual of Nearctic Diptera. Vol. 1. Agriculture Canada Monograph 27, 674 pp.

Eldredge, N. and S.J. Gould. 1972. Punctuated equilibria: an alternative to phyletic gradualism. pp. 82-115 *In* T.J. M. Schopf, ed., Models in Paleobiology. Freeman, Cooper and Co., San Francisco, California.

Evenhuis, N.L. 1994. Catalogue of the fossil flies of the world (Insecta: Diptera). Backhuys Publishers, Leiden, The Netherlands, 600 pp.

Ewing, H.E. 1937. Arachnida. Order Acarina. pp. 56-62, *In* Insects and arachnids from Canadian amber, by F.M. Carpenter, J.W. Folsom, E.O. Essig, A.C. Kinsey, C.T. Brues, M.W. Boesel, and H.E. Ewing. U. of Toronto Studies, Geological Series No. 40:7-62.

Farris, J.S. 1986. On the boundaries of phylogenetic systematics. Cladistics 2:14-27.

Felippe-Bauer, M.L. and P.G. Bauer. 1990. Sensilla ampullacea on the antennae of *Culicoides paraensis* (Goeldi, 1905) with notes on other *Culicoides* (Diptera: Ceratopogonidae). Memorias do Instituto Oswaldo Cruz 85:235-237.

Felippe-Bauer, M.L., P.G. Bauer, and F.C. Silva Filho. 1989. Scanning electron microscopy of the antennal sensilla in female *Culicoides paraensis* (Diptera: Ceratopogonidae). Memorias do Instituto Oswaldo Cruz 84:463-469.

Folinsbee, R.E., H. Baadsgaard, G.L. Cumming and J. Nascimbene. 1964. Radiometric dating of the Bearpaw Sea. Bulletin of the American Association of Petroleum Geologists 48:525.

Förster, B. 1891. Die Insekten des "Plattigen Steinmergels" von Brunstatt. Abhandlungen zur Geologischen Specialkarte Elsass-Lotheringen 3:333-594.

Frazier, J.L. 1985. Nervous system: sensory system. Chapter 8 *In* Fundamental of Insect Physiology, ed. M.S. Blum, John Wiley and Sons, New York.

Gad, A.M. 1951. The head-capsule and mouth-parts in the Ceratopogonidae (Diptera - Nematocera). Bulletin de la Société Fouad 1er d'Entomologie 35:17-75.

Gagné, R.J. 1977. Cecidomyiidae (Diptera) from Canadian amber. Proceedings of the Entomological Society of Washington 79:57-62.

Glukhova, V.M. 1981. A comparative morphological review of the mouthparts of females and the males in the subfamilies Dasyheleinae and Forcipomyiinae (Diptera, Ceratopogonidae). Entomologischeskoe Obozrenie 60:62-76 (in Russian, translated in Entomological Review 60:59-72).

Glukhova, V.M. 1982. On the structure of mouth parts in bloodsucking midges of the subgenus *Trithecoides* of the genus *Culicoides* (Ceratopogonidae) (in Russian). Parasitologiya 16:155-159.

Glukhova, V.M. 1989. Blood-sucking midges of the genera *Culicoides* and *Forcipomyia* (Ceratopogonidae) (in Russian). Fauna of the USSR, 139. Vol. 3(5a). 408 pp.

Goetghebuer, M. 1923. Cératopogonines de Belgique parasites accidentels de l'homme. Bulletin de la Société Entomologique de Belgique 5:34-37.

Grimaldi, D.A. 1990. Chapter 9. Diptera. pp. 164-183 *In* Insects from the Santana Formation, Lower Cretaceous, of Brazil. Bulletin of the American Museum of Natural History 195, 191 pp.

Grimaldi, D.A. 1993a. Forever in amber. Natural History 6/93:59-61.

Grimaldi, D.A. 1993b. The care and study of fossiliferous amber. Curator 36:31-49.

Grimaldi, D., C.W. Beck, and J.J. Boon. 1989. Occurrence, chemical characteristics, and paleontology of the fossil resins from New Jersey. American Museum Novitates 2948, 27 pp.

Grogan, W.L. and A. Borkent. 1992. *Sinhalohelea*, a new genus of predaceous midge from Sri Lanka (Diptera: Ceratopogonidae). Proceedings of the Entomological Society of Washington 94:314-319.

Grogan, W.L. and R. Szadziewski. 1988. A new biting midge from Upper Cretaceous (Cenomanian) amber of New Jersey (Diptera: Ceratopogonidae). Journal of Paleontology 62:808-812.

Grogan, W.L. and W.W. Wirth. 1978. Description of the male of *Neurobezzia granulosa* (Wirth) (Diptera: Ceratopogonidae). Proceedings of the Entomological Society of Washington 80:548-552.

Grogan, W.L. and W.W. Wirth. 1979a. The North American predaceous midges of the genus *Palpomyia* Meigen (Diptera: Ceratopogonidae). Memoirs of the Entomological Society of Washington 8, 125 pp.

Grogan, W.L. and W.W. Wirth. 1979b. A new neotropical genus of predaceous midges, with a key to the genera of Heteromyiini (Diptera: Ceratopogonidae). Proceedings of the Entomological Society of Washington 81:51-59.

Grogan, W.L. and W.W. Wirth. 1980. *Nannohelea*, a new genus of biting midges of the tribe Ceratopogonini, related to *Baeohelea* Wirth and Blanton (Diptera: Ceratopogonidae). Journal of the Kansas Entomological Society 53:373-385.

Hamilton, K.G.A. 1971. A remarkable fossil homopteran from Canadian Cretaceous amber representing a new family. Canadian Entomologist 103:943-946.

Harbach, R.E. and K.L. Knight. 1980. Taxonomists' glossary of mosquito anatomy. Plexus Publishing, Marlton, New Jersey. 415 pp.

Heie, O.E. and E.M. Pike. 1992. New aphids in Cretaceous amber from Alberta (Insecta, Homoptera). Canadian Entomologist 124:1027-1053.

Hennig, W. 1966. The Diptera fauna of New Zealand as a problem in systematics and zoogeography. Pacific Insects Monograph 9:1-81.

Heyden, L., von. 1870. Fossile Dipteren aus der Braunkohle von Rott im Siebengebirge. Palaeontographica 17:237-266, pls. 44-45.

Hong, Y.-C. 1981. Eocene fossil Diptera in amber in Fushun coalfield [in Chinese, with English summary]. Geological Publishing House, Peking, 166 pp.

Hurd, P.D., R.F. Smith and J.W. Durham. 1962. The fossiliferous amber of Chiapas, Mexico. Ciencia (Mexico) 21:107-118.

Issayev, A. 1983. Results of a comparative karyological analysis of Ceratopogoninae and Palpomyiinae (in Russian). pp. 56-58 *In* Diptera (Insecta), their systematics, geographic distribution and ecology. USSR Academy of Sciences, Zoological Institute, Moscow. 156 pp.

Jamnback, H. 1965. The *Culicoides* of New York State (Diptera: Ceratopogonidae). New York State Museum and Science Service Bulletin Number 399, viii + 154 pp.

Jarzembowski, E. and A. Ross. 1994. Progressive palaeontology. Antenna 18(3):123-126.

Jarzen, D.M. 1982. Palynology of Dinosaur Provincial Park (Campanian) Alberta. Syllogeus 38:1-69.

Jell, P.A. and P.M. Duncan. 1986. Invertebrates, mainly insects, from the freshwater, Lower Cretaceous, Koonwarra fossil bed (Korumburra Group), South Gippsland, Victoria. Memoir of the Association of Australasian Palaeontologists 3:111-205.

Jobling, B. 1928. The structure of the head and mouth parts in *Culicoides pulicaris*, L. (Diptera Nematocera). Bulletin of Entomological Research 18:211-236, pls. 9-12.

Johannsen, O.A. 1943. A generic synopsis of the Ceratopogonidae (Heleidae) of the Americas, a bibliography, and a list of the North American species. Annals of the Entomological Society of America 36:763-791.

Kalugina, N.S. 1985. *In* N.S. Kalugina and V.G. Kovalev. Jurassic Diptera of Siberia (in Russian). USSR Academy of Sciences, Moscow. 199 pp.

Kalugina, N.S 1986. pp.112-125, *In* Insects in the Early Cretaceous ecosystems of the West Mongolia (in Russian). The joint Soviet-Mongolian Palaeontological Expedition. Academy of Sciences of the USSR, Paleontological Institute. Transactions. Volume 28, 214 pp.

Kalugina, N.S. 1991. New Mesozoic Simuliidae and Leptoconopidae and blood-sucking origin in lower Dipterans [in Russian, with English summary]. Paleontologisceskij Zhurnal 1991(1):69-80, pl. 6. English translation in Paleont. J. 25:66-77 (1992).

Keilbach, R. 1982. Bibliographie und Liste der Arten tierischer Einschlüsse in fossilen Harzen sowie ihrer Aufbewahrungsorte. Teil 1,2. Deutsche Entomologische Zeitschrift (N.F) 29:129-286, 301-491.

Keilin, D. 1921. On the life-history of *Dasyhelea obscura*, Winnertz (Diptera, Nematocera, Ceratopogonidae), with some remarks on the parasites and hereditary bacterian symbiont of this midge. Annals and Magazine of Natural History 9 (Series 9):576-590, pls. 19-20.

Kellogg, F.E. 1970. Water vapour and carbon dioxide receptors in *Aedes aegypti*. Journal of Insect Physiology 16:99-108.

Kettle, D.S. 1955. Sex ratios among British *Culicoides*. Proceedings of the Entomological Society of London (A) 30:70-72.

Kieffer, J.J. 1899. Description d'un nouveau genre et tableau des genres européens de la famille des Chironomides (Dipt.). Bulletin de la Société Entomologique de France 1899:66-70.

Kieffer, J.J. 1906. Diptera. Fam. Chironomidae. Fasc. 42, 78 pp., 4 pls. *In* Wytsman, P., ed., Genera Insectorum. Bruxelles.

Kieffer, J.J. 1911a. Nouvelles descriptions de Chironomides obtenus d'éclosion. Bulletin de la Société d'Histoire Naturelle de Metz 27:1-60.

Kieffer, J.J. 1911b. XI. Description de nouveaux Chironomides de l'Indian Museum de Calcutta. Records of the Indian Museum 6:113-177, pls. VI-VII.

Kieffer, J.J. 1913. Chironomidae et Cecidomyidae. pp. 1-43 *In* Voyage de Ch. Alluaud et R. Jeannel en Afrique orientale. Insectes Diptères [= Vol. 5] (eds. C.A. Alluaud and R. Jeannel), 351 pp, Paris.

Kieffer, J.J. 1917. Chironomides d'Amérique conservés au Musée National Hongrois de Budapest. Annales Musei Nationalis Hungarici 15:292-364.

Kieffer, J.J. 1925. Diptères (Nématocères piqueurs): Chironomidae Ceratopogoninae. Faune de France 11:1-139.

Klimaszewski, J. and D.K. McE. Kevan. 1986. A new lacewing-fly (Neuroptera: Planipennia) from Canadian Cretaceous amber, with an analysis of its forewing characters. Entomological News 97:124-132.

Klotz, O.J. 1892. Survey of amber locations. pp. 18-27. *In* Annual Report of the Department of the Interior for the Year 1891 (No. 13-1891), Part II. Ottawa, Ontario, Canada.

Knausenberger, W.I. 1987. Contributions to the autecology and ecosystematics of immature Ceratopogonidae (Diptera), with emphasis on the tribes Heteromyiini and Sphaeromiini in the Middle Atlantic United States. Ph.D. Dissertation, Virginia Polytechnic Institute and State University, Blacksburg, Virginia, xxix + 729 pp.

Krzeminska, E. and W. Krzeminski. 1992. Les Fantomes de l'Ambre. Insectes fossiles dans l'ambre de la Baltique. Musée d'histoire naturelle de Neuchatel, Suisse, 142 pp.

Krzeminski, W. and H.J. Teskey. 1987. New taxa of Limoniidae (Diptera: Nematocera) from Canadian amber. Canadian Entomologist 119:887-892.

Labandeira, C.C. and J.J. Sepkoski. 1993. Insect diversity in the fossil record. Science 261:310-315.

Lambert, J.B., J.S. Frye and G.O. Poinar. 1990. Analysis of North American amber by Carbon-13 NMR spectroscopy. Geoarchaeology 5:43-52.

Langenheim, J.H. 1969. Amber: a botanical inquiry. Science 163:1157-1169.

Langenheim, J.H. 1990. Plant resins. American Scientist 78:16-24.

Langenheim, J.H. and C.W. Beck. 1968. Catalogue of infrared spectra of fossil resins (ambers). I. North and South America. Botanical Museum Leaflets, Harvard University 22:65-120.

Langenheim, R.L., C.J. Smiley and J. Gray. 1960. Cretaceous amber from the arctic coastal plain of Alaska. Bulletin of the Geological Society of America 71:1345-1356.

Larsson, S.G. 1978. Baltic amber - a palaeobiological study. Entomonograph 1, 192 pp.

Latreille, P.A. 1809. Genera crustaceorum et insectorum secundum ordinem naturalem in familias disposita, iconibus exemplisque plurimis explicata. Vol. 4, Paris and Strasbourg, 399 pp., 4 pls.

Lee, D.J., E.J. Reye, and A.L. Dyce. 1962. "Sandflies" as possible vectors of disease in domesticated animals in Australia. Proceedings of the Linnean Society of New South Wales 87:364-376.

Lewis, D.J. 1975. Functional morphology of the mouth parts in New World phlebotomine sandflies (Diptera: Psychodidae). Transactions of the Royal Entomological Society of London 126:497-532.

Lewis, D.J. and R.P. Lane. 1976. A taxonomic review of *Phlebotomus* (*Idiophlebotomus*) (Psychodidae). Systematic Entomology 1:53-60.

Linley, J.R. 1968. Autogeny and polymorphism for wing length in *Leptoconops becquaerti* (Kieff.) (Diptera: Ceratopogonidae). Journal of Medical Entomology 5:53-66.

Linley, J.R. 1976. Biting midges of mangrove swamps and salt marshes. pp. 335-376 *In* Marine insects, L. Cheng ed. xii + 581 pp., Elsevier.

Linley, J.R. and J.B. Davies. 1971. Sand flies and tourism in Florida and the Bahamas and Caribbean area. Journal of Economic Entomology 64:264-278.

Linnaeus, C. 1767. Systema naturae per regna tria naturae. Ed. 12 (rev.), Vol. 2, Pt. 2, pp. 533-1327. Homiae [= Stockholm].

Loew, H. 1850. Ueber den Bernstein und die Bernsteinfauna. Program König. Realschule Meseritz, 1850, 1-48.

Macfie, J.W.S. 1940. The genera of Ceratopogonidae. Annals of Tropical Medicine and Parasitology 34:13-30.

MacKay, M.R. 1970. Lepidoptera in Cretaceous amber. Science 167:379-380.

Magnon, G.J., D.V. Hagan, D.L. Kline, and J.R. Linley. 1990. Habitat characteristics and phenology of larval *Culicoides* spp. (Diptera: Ceratopogonidae) from a coastal Georgia salt marsh. Environmental Entomology 19:1068-1074.

Marincovich, L., E.M. Brouwers, D.M. Hopkins, and M.C. McKenna. 1990. Late Mesozoic and Cenozoic paleogeographic and paleoclimatic history of the Arctic Ocean basin, based on shallow-water marine faunas and terrestrial vertebrates. pp. 403-426 *In* The Geology of North America, Vol. 50, The Arctic Ocean Region. Geological Society of America.

Marx, J.L. 1978. Warm-blooded dinosaurs: evidence pro and con. Science 199:1424-1426.

Matayoshi, S., S. Noda, and A. Sato. 1985. Ecological study of *Leptoconops (Leptoconops) nipponensis oshimaensis* (Diptera, Ceratopogonidae) at Katoku, Amami-oshima Island, Japan. Japanese Journal of Sanitary Zoology 36:219-225.

Mayer, K. 1955. Der Parasitismus der Heleiden (Ceratopogoniden) Dipt. Zeitschrift für Angewandte Zoologie 1:95-107.

McAlpine, J.F. 1973. A fossil ironomyiid fly from Canadian amber (Diptera: Ironomyiidae). Canadian Entomologist 105: 105-111.

McAlpine, J.F. 1981. 2. Morphology and terminology - adults. pp. 9-63 in Manual of Nearctic Diptera. Volume 1. Agriculture Canada Monograph 27, 674 pp.

McAlpine, J.F. and J.E.H. Martin. 1966. Systematics of Sciadoceridae and relatives with description of two new genera and species from Canadian amber and erection of family Ironomyiidae (Diptera: Phoroidea). Canadian Entomologist 98:527-544.

McAlpine, J.F. and J.E.H. Martin. 1969. Canadian amber - a palaeontological treasure-chest. Canadian Entomologist 101:819-838.

McIver, S. and C. Charlton. 1970. Studies on the sense organs on the palps of selected culicine mosquitoes. Canadian Journal of Zoology 48:293-295.

McKeever, S. 1986. Mouthparts of the four North American *Corethrella* species (Diptera: Chaoboridae), with detailed study of *C. appendiculata*. Journal of Medical Entomology 23:502-512.

McKeever, S. and J.M. Pound. 1979. The mouthparts of female *Corethrella brakeleyi* and *C. wirthi* (Diptera: Chaoboridae). Journal of Morphology 161:157-168.

McKeever, S., M.D. Wright, and D.V. Hagan. 1988. Mouthparts of females of four *Culicoides* species (Diptera: Ceratopogonidae). Annals of the Entomological Society of America 81:332-341.

McKeever, S., D.V. Hagan, and W.L. Grogan. 1991. Comparative study of mouthparts of ten species of predaceous midges of the tribe Ceratopogonini (Diptera: Ceratopogonidae). Annals of the Entomological Society of America 84:93-106.

Meigen, J.W. 1803. Versuch einer neuen Gattungseintheilung der europäischen zweiflügligen Insekten. Magazin für Insektenkunde 2:259-281.

Meigen, J.W. 1818. Systematische Beschreibung der bekannten europäischen zweiflügeligen Insekten. Vol. 1, xxxvi + 333 pp., pls. 1-11. Aachen.

Meiswinkel, R. 1992. Afrotropical *Culicoides*: *C.* (*Avaritia*) *loxodontis* sp. nov., a new member of the *imicola* group (Diptera: Ceratopogonidae) associated with the African Elephant in the Kruger National Park, South Africa. Onderstepoort Journal of Veterinary Research 59:145-160.

Mercer, K.L. and S.B. McIver. 1973. Sensilla on the palps of selected Blackflies (Diptera: Simuliidae). Journal of Medical Entomology 10:236-239.

Messaddeq, N., M. Fabre and M. Kremer. 1989. Étude au microscope électronique a balayage des organes sensoriels de *Culicoides nubeculosus* (Diptére: Ceratopogonidé). Annales de Parasitologie Humaine et Comparee 64:224-237.

Meunier, F. 1904. Monographie des Cecidomyidae, des Sciaridae, des Mycetophilidae et des Chironomidae de l'ambre de la Baltique. Brussels. 264 pp. (pp. 82-264 predate the part of the journal version which appeared in Annales de la Société Scientifique de Bruxelles 28:93-275).

Meunier, F. 1912. Quelques Chironomidae du copal récent de Zanzibar et de Madagascar (Dipt.). Bulletin de la Société Entomologique de France 1912:361-366.

Meunier, F. 1920. Quelques insectes de l'Aquitainen de Rott, Sept-Monts (Presse rhénane) [concl.]. Verslag Koninklijke Akademie van Wetenschappen te Amsterdam (II) 28:1215-1222. (Also published 1920, Proc. Koninklijke Akademie van Wetenschappen te Amsterdam 22:891-898).

Montgomery de Merette, L. 1984. L'Ambre de St. Domingue. Monde & Mineraux 10:36-37, 40-41.

Nielsen, A. 1951. Contributions to the metamorphosis and biology of the genus *Atrichopogon* Kieffer. Kongelige Danske Videnskabernes Selskab, Biologiske Skrifter 6(6):1-95, pls. 1-2.

Norris, G. 1975. Evolution of the Cretaceous terrestrial palynoflora in western Canada. The Geological Association of Canada Special Paper Number 13:333-364.

Norris, G. 1985. Palynology and British Purbeck facies. Geological Magazine 122:187-190.

Obradovich, J.D. 1993. A Cretaceous time scale. pp. 379-396 *In* Evolution of the Western Interior Basin. Geological Association of Canada, Special Paper 39, eds. W.G.E. Caldwell and E.G Kauffman.

Palmer, A.R. 1957. Miocene arthropods from the Mojave Desert, California. Geological Survey Professional Paper, 294-G:237-277, pl. 34.

Pellegrino, C.R. 1985. Dinosaur Capsule. Omni 7(4):38-40, 114-115.

Penny, J.S. 1947. Studies on the conifers of the Magothy flora. American Journal of Botany 34:281-295.

Peterson, B.V. 1975. A new Cretaceous bibionid from Canadian amber (Diptera: Bibionidae). Canadian Entomologist 107:711-715.

Petrunkevitch, A. 1956. *Eohelea stridulans*, a striking example of paramorphism in a Baltic amber gnat. Science 123:675.

Petrunkevitch, A. 1957. *Eohelea stridulans*, n.gen., n.sp., a striking example of paramorphism in an amber biting- midge. Journal of Paleontology 31:208-214.

Pierce, W.D. 1959. Fossil arthropods of California. No. 22. A progress report on the nodule studies. Bulletin of the Southern California Academy of Sciences 58:72-78.

Pierce, W.D. 1966. Fossil arthropods of California. 29. Silicified Miocene pupae of ceratopogonid flies. Bulletin of the Southern California Academy of Sciences 65(2):81-98.

Pike, E. 1991. How old is the Cedar Lake amber? pp. 56, Abstract, *In* Joint Annual Meeting. Entomological Society of Canada, Société d'Entomologie du Québec, Entomological Society of Ontario. Oct. 20-23, 1991, 80 pp.

Pike, E.M. 1994. Historical changes in insect community structure as indicated by hexapods of Upper Cretaceous Alberta (Grassy Lake) amber. Canadian Entomologist 126:695-702.

Poinar, G.O. 1992. Life in amber. Stanford University Press, Stanford, California. xiii + 350 pp.

Poinar, G.O. and J. Haverkamp. 1985. Use of pyrolysis mass spectrometry in the identification of amber samples. Journal of Baltic Studies 16:210-221.

Poinar, G.O., E.M. Pike and G.W. Krantz. 1993. Animal - animal parasitism. Nature 361:307-308.

Poinar, G.O., A. Acra and F. Acra. 1994. Animal-animal parasitism in Lebanese amber. Medical Science Research 22:159.

Rees, D.M., P.G. Lawyer and R.N. Winget. 1971. Colonization of *Leptoconops kerteszi* Kieffer by anautogenous and autogenous reproduction (Diptera: Ceratopogonidae). Journal of Medical Entomology 8:266-271.

Remm, H. 1975. On the classification of the biting midges (Diptera, Ceratopogonidae) (in Russian). Parasitologiya 9:393-397.

Remm, H. 1976. Midges (Diptera, Ceratopogonidae) from the Upper Cretaceous fossil resins of the Khatanga depression (in Russian). Paleontologisceskij Zhurnal 3:107-116.

Retallack, G.J. and D.L. Dilcher. 1986. Cretaceous Angiosperm invasion of North America. Cretaceous Research 7:227-252.

Richards, W.R. 1966. Systematics of fossil aphids from Canadian amber (Homoptera: Aphididae). Canadian Entomologist 98:746-760.

Rioux, J.A., S. Descous, J.J. Corre, H. Croset and S. Tour. 1968. Ecologie de *Leptoconops irritans* Noe, 1905 en Moyenne Camargue. Localisation et dynamique des biotopes larvaires. La Terre et la Vie 22:458-469.

Rohdendorf, B.B. and V.V. Zherichin. 1974. Paleontology and the preservation of nature (in Russian). Privoda 5:82-91.

Romer, A.S. 1966. Vertebrate Paleontology. University of Chicago Press. Chicago and London. ix + 468 pp.

Rowley, W.A. and M. Cornford. 1972. Scanning electron microscopy of the pit of the maxillary palp of selected species of *Culicoides*. Canadian Journal of Zoology 50:1207-1210.

Russell, D.A. 1977. A vanished world, the dinosaurs of western Canada. National Museum of Natural Sciences, Ottawa, Canada, 142 pp.

Russell, D.A. 1989. An odyssey in time: The Dinosaurs of North America. University of Toronto Press, Toronto, Canada, 240 pp.

Saether, O.A. 1970. Nearctic and Palaearctic *Chaoborus* (Diptera: Chaoboridae). Fisheries Research Board of Canada, Bulletin 174, vii + 57 pp.

Saether, O.A. 1977. Female genitalia in Chironomidae and other Nematocera: morphology, phylogenies, keys. Bulletin of the Fisheries Research Board of Canada 197. 209 pp.

Saether, O.A. 1979a. Underlying synapomorphies and anagenetic analysis. Zoologica Scripta 8:305-312.

Saether, O.A. 1979b. Underliggende synapomorfi og enestaende innvendig parallellisme belyst ved kesempler fra Chironomidae og Chaoboridae [in Norwegian, with English summary]. Entomologisk Tidskrift 100:173-180.

Saether, O.A. 1980. Glossary of chironomid morphology terminology (Diptera: Chironomidae). Entomologica Scandinavica Supplement 14:1-51.

Saether, O.A. 1983. The canalized evolutionary potential: inconsistencies in phylogenetic reasoning. Systematic Zoology 32:343-359.

Saether, O.A. 1986. The myth of objectivity - post-Hennigian deviations. Cladistics 2:1-13.

Saether, O.A. 1988. On the limitation of parsimony in phylogenetic analysis. Quaestiones Entomologicae 24:45-50.

Saether, O.A. and G.A. Halvorsen. 1981. Diagnoses of *Tvetenia* Kieff. emend., *Dratnalia* n.gen., and *Eukiefferiella* Thien. emend., with a phylogeny of the *Cardiocladius* group (Diptera: Chironomidae). Entomologica Scandinavica Supplement 15:269-285.

Santos Abreu, E. 1918. Ensayo de una Monografia de los Tendipedidos de las Islas Canarias. Memorias de la Real Academia de Ciencias y Artes de Barcelona 14(2):159-326, 1 pl.

Schawaller, W. 1991. The first Mesozoic Pseudoscorpion, from Cretaceous Canadian amber. Palaeontology 34:971-976.

Schlee, D. 1984. Notizen über einige Bernsteine und Kopale aus aller Welt. Stuttgarter Beiträge zur naturkunde. Series C. 18:29-62.

Schlee, D. and H.G. Dietrich. 1970. Insektenführender Bernstein aus der Unterkreide des Libanon. Neues Jahrbuch für Geologie und Paläontologie 1970 (1):40-50.

Schlee, D. and W. Glöckner. 1978. Bernstein, Bernsteine und Bernstein-Fossilien. Stuttgarter Beitrage zur Naturkunde, Ser. C, 8:1-72.

Schlüter, T. 1978. Zur Systematik und Palökologie harzkonservierter Arthropoda einer Taphozönose aus dem Cenomanium von NW-Frankreich. Berliner Geowissenschaftliche Abhandlungen, Reihe A 9:1-150.

Schlüter, T. 1983. A fossiliferous resin from the Cenomanian of the Paris and Aquitanian Basins of Northwestern France. Cretaceous Research 4:265-269.

Schlüter, T. and F. von Gnielinski 1987. The East African copal. Its geologic, stratigraphic, palaeontologic significance and comparison with other fossil resins of similar age. National Museums of Tanzania, Occasional Paper 8, 32 pp.

Schumann, H. and H. Wendt. 1989. Zur Kenntnis der tierischen Inklusen des Sächsischen Bernsteins. Deutsche Entomologische Zeitschrift 36:33-44.

Sereno, P.C. and R. Chenggang. 1992. Early evolution of avian flight and perching: new evidence from the lower Cretaceous of China. Science 255:845-848.

Shedrinsky, A.M., D. Grimaldi, T.P. Wampler and N.S. Baer. 1991. Amber and copal: pyrolysis gas chromatographic (PyGC) studies of provenance. Wiener Berichte über Naturwissenschaft in der Kunst 6/7/8:37-63.

Sheehan, P.M. and D.E. Fastovsky. 1992. Major extinctions of land-dwelling vertebrates at the Cretaceous-Tertiary boundary, eastern Montana. Geology 20:556-560.

Sinclair, B.J. 1992a. A new species of *Trichothaumalea* (Diptera: Thaumaleidae) from eastern North America and a discussion of male genitalic homologies. Canadian Entomologist 124: 491-499.

Sinclair, B.J. 1992b. A phylogenetic interpretation of the Brachycera (Diptera) based on the larval mandible and associated mouthpart structures. Systematic Entomology 17:233-252.

Sinclair, B.J., J.M. Cumming, and D.M. Wood. 1994. Homology and phylogenetic implications of male genitalia in Diptera - Lower Brachycera. Entomologica Scandinavica 24:407-432.

Skuse, F.A.A. 1889. Diptera of Australia. Part VI. - The Chironomidae. Proceedings of the Linnean Society of New South Wales 4:215-311, pls. XI-XIV,XIV bis.

Smee, L. 1966. A revision of the subfamily Leptoconopinae Noé (Diptera: Ceratopogonidae) in Australasia. Australian Journal of Zoology 14:993-1025.

Smith, A.G., A.M. Hurley, and J.C. Briden. 1981. Phanerozoic Paleocontinental World Maps. Cambridge University Press, Cambridge. 102 pp.

Smith, L.M. and H. Lowe. 1948. The Black Gnats of California. Hilgardia 18:157-183.

Spicer, R.A. and J.T. Parrish. 1987. Paleobotanical evidence for cool north polar climates in middle Cretaceous (Albian-Cenomanian) time. Geology 14:703-706.

Statz, G. 1941. Plagegeifter aus dem rheinifchen Braunkohlenwalde. Rheinischer Naturfreund 5(1):1-16.

Statz, G. 1944. Neue Dipteren (Nematocera) aus dem Oberoligozän von Rott. VI. Familie: Tendipedidae (Zuck-oder Schwarmücken). VII. Familie: Heleidae (Ceratopogonidae), (Gnitzen). VIII. Familie: Lycoriidae (Trauermücken). Palaeontographica Abt. A, 95:123-188, 17 pls.

Stockey, R.A. 1982. The Araucariaceae: an evolutionary perspective. Review of Palaeobotany and Palynology 37:133-154.

Sutcliffe, J.F. 1987. Distance orientation of biting flies to their hosts. Insect Science and its Application 8:611-616.

Sutcliffe, J.F. and P.D. Deepan. 1988. Anatomy and function of the mouthparts of the biting midge, *Culicoides sanguisuga* (Diptera: Ceratopogonidae). Journal of Morphology 198: 353-365.

Sutcliffe, J.F. and S.B. McIver. 1984. Mechanics of blood-feeding in Black Flies (Diptera, Simuliidae). Journal of Morphology 180:125-144.

Sutcliffe, J.F., J.L. Shipp, and E.G. Kokko. 1987. Ultrastructure of the palpal bulb sensilla of the Black Fly *Simulium arcticum* (Diptera: Simuliidae). Journal of Medical Entomology 24:324-331.

Szadziewski, R. 1984. Niezwykle narzady strydulacyjne u eocenskich muchowek z rodziny Ceratopogonidae (Diptera). Wiad. Entomol. 5(1-2):37-40.

Szadziewski, R. 1985. Biting midges of the genus *Eohelea* Petrunkevitch (Insecta, Diptera, Ceratopogonidae) from Baltic amber (in the collection of the Museum of the Earth). Prace Muzeum Ziemi 37:123-130.

Szadziewski, R. 1988. Biting midges (Diptera, Ceratopogonidae) from Baltic amber. Polskie Pismo Entomologiczne 58:3-283.

Szadziewski, R. 1990a. Biting midges (Insecta: Diptera: Ceratopogonidae) from Sakhalin amber. Prace Muzeum Ziemi 41:77-81.

Szadziewski, R. 1990b. The oldest fossil biting midges /Ceratopogonidae/. pp. 231, Second International Congress of Dipterology, Bratislava, Czechoslovakia. Abstract Volume, 324 pp.

Szadziewski, R. 1993. Biting midges (Diptera, Ceratopogonidae) from Miocene Saxonian amber. Acta zoologica cracoviensia 35:603-656.

Szadziewski, R. and W.L. Grogan. 1994a. Biting midges from Dominican amber. I. A new fossil species of *Baeodasymyia* (Diptera: Ceratopogonidae). Proceedings of the Entomological Society of Washington 96:219-229.

Szadziewski, R. and W.L. Grogan. 1994b. The biting midge fauna in Dominican amber (Diptera: Ceratopogonidae). Third International Congress of Dipterology, Abstract Volume, pp. 220-221.

Szadziewski, R. and T. Schlüter. 1992. Biting midges (Diptera: Ceratopogonidae) from Upper Cretaceous (Cenomanian) amber of France. Annales de la Société Entomologique de France 28:73-81.

Teskey, H.J. 1971. A new soldier fly from Canadian amber (Diptera: Stratiomyidae). Canadian Entomologist 103:1659-1661.

Theischinger, G. 1986. Australian Thaumaleidae (Insecta: Diptera). Records of the Australian Museum 38:291-317.

Thomas, D.K. and E.C. Bland (eds.). 1980. A cold look at the warm-blooded dinosaurs. Westview Press.

Tokunaga, M. 1966. Biting midges of the Palpomyiinae from New Guinea (Diptera: Ceratopogonidae). Pacific Insects 8:101-152.

Tyrrell, J.B. 1892. Report on north-western Manitoba with portions on the adjacent districts of Assiniboia and Saskatchewan. Annual Report. Geological Survey of Canada 5(n.s.), Part 1, Report E.

Van Valen, L.M. 1984. Catastrophes, expectations, and the evidence. Paleobiology 10:121-137.

Walker, T.L. 1934. Chemawinite or Canadian amber. University of Toronto Studies, Geological Series 36:5-10, pls. 1-2.
Watrous, L.E. and Q.D. Wheeler. 1981. The outgroup comparison method of character analysis. Systematic Zoology 30:1-11.
Wenk, P. 1953. Der Kopf von *Ctenocephalus canis* (Curt.) (Aphaniptera). Zoologische Jahrbücher. Anatomie 73:103-164.
Westwood, J.O. 1840. Order XIII. Diptera Aristotle (Antliata Fabricius. Halteriptera Clairv.). Pp. 125-154. *In his* An introduction to the modern classification of insects; founded on the natural habits and corresponding organisation of the different families. Synopsis of the genera of British insects, 158 pp. London.
Westwood, J.O. 1854. 5. Contributions to fossil entomology. Quarterly Journal of the Geological Society of London 10:378-396, pls. 14-18.
Whalley, P. 1987. Insects and Cretaceous mass extinction. Nature 327:562.
Wiley, E.O. 1981. Phylogenetics. The theory and practice of phylogenetic systematics. J. Wiley and Sons, New York. xv + 439 pp.
Willassen, E. 1983. Notes on the chironomid (Diptera, Chironomidae) “ocelli”. Zoologica Scripta 12:221-223.
Williams, F.X. 1944. Biological studies in Hawaiian water-loving insects. Part III. Diptera or flies. D. Culicidae, Chironomidae and Ceratopogonidae. Proceedings of the Hawaiian Entomological Society 12:149-180.
Wilson, E.O. 1985. Ants from the Cretaceous and Eocene amber of North America. Psyche 92:205-216.
Wirth, W.W. 1951. A new biting midge of the genus *Leptoconops* from Florida, with new records of other American species. Proceedings of the Entomological Society of Washington 53:281-284.
Wirth, W.W. 1952. The Heleidae of California. University of California Publications in Entomology 9:95-266.
Wirth, W.W. 1977. A review of the pathogens and parasites of the biting midges (Diptera: Ceratopogonidae). Journal of the Washington Academy of Sciences 67:60-75.
Wirth, W.W. 1981. *Paradasyhelea harrisoni* n.sp. from the Auckland Islands, and additional records of subantarctic Ceratopogonidae. New Zealand Journal of Zoology 8:383-386.
Wirth, W.W. 1991. Notes and corrections on stick-ticks, Neotropical parasitic midges of the *Forcipomyia* subgenus *Microhelea* (Diptera: Ceratopogonidae). Florida Entomologist 74:122-128.
Wirth, W.W. and W.R. Atchley. 1973. A review of the North American *Leptoconops* (Diptera: Ceratopogonidae). Graduate Studies Texas Tech University (5), 57 pp.
Wirth, W.W. and F.S. Blanton. 1953. Studies in Panama *Culicoides* (Diptera: Heleidae): I, Descriptions of six new species. Journal of the Washington Academy of Sciences 43:69-77.
Wirth, W.W. and F.S. Blanton. 1959. The genus *Paradasyhelea* Macfie, with descriptions of two new species from eastern Australia (Diptera: Ceratopogonidae). Bulletin of the Brooklyn Entomological Society 54:114-121.
Wirth, W.W. and F.S. Blanton. 1969. North America *Culicoides* of the *pulicaris* group (Diptera: Ceratopogonidae). Florida Entomologist 52:207-243.
Wirth, W.W. and F.S. Blanton. 1970. New genera of Neotropical Ceratopogonidae (Diptera). Florida Entomologist 53:7-14.
Wirth, W.W. and F.S. Blanton. 1973. A review of the maruins or biting midges of

the genus *Culicoides* (Diptera: Ceratopogonidae) in the Amazon Basin. Amazoniana 4:405-470.

Wirth, W.W. and W.L. Grogan. 1988. The predaceous midges of the world (Diptera: Ceratopogonidae; Tribe Ceratopogonini). Flora and Fauna Handbook 4, xv + 160 pp., E.J. Brill.

Wirth, W.W. and A.A. Hubert. 1962. The species of *Culicoides* related to *piliferus* Root and Hoffman in eastern North America (Diptera, Ceratopogonidae). Annals of the Entomological Society of America 55:182-195.

Wirth, W.W. and A.A. Hubert. 1989. The *Culicoides* of Southeast Asia (Diptera: Ceratopogonidae). Memoirs of the American Entomological Institute 44, 508 pp.

Wirth, W.W. and D.J. Lee. 1958. Australasian Ceratopogonidae (Diptera, Nematocera). Part VIII: A new genus from Western Australia attacking man. Proceedings of the Linnean Society of New South Wales 83:337-339.

Wirth, W.W. and D.J. Lee. 1959. The genus *Paradasyhelea* Macfie, with descriptions of two new species from eastern Australia (Diptera: Ceratopogonidae). Bulletin of the Brooklyn Entomological Society 54:114-121.

Wirth, W.W. and S. Navai. 1978. Terminology of some antennal sensory organs of *Culicoides* biting midges (Diptera: Ceratopogonidae). Journal of Medical Entomology 15:43-49.

Wirth, W.W. and N.C. Ratanaworabhan. 1971. *Ceratoculicoides*, a new genus related to *Ceratopogon* Meigen (Diptera: Ceratopogonidae). Proceedings of the Entomological Society of Washington 73:170-177.

Wirth, W.W. and N.C. Ratanaworabhan. 1993. The Oriental biting midges of the genus *Atrichopogon* related to *Dolichohelea polita* Edwards (Diptera: Ceratopogonidae). Oriental Insects 27:317-334.

Wirth, W.W., N.C. Ratanaworabhan, and F.S. Blanton. 1974. Synopsis of the genera of Ceratopogonidae (Diptera). Annales de Parasitologie Humaine et Comparee 49:595-613.

Wolfe, J.A. and G.R. Upchurch. 1986. Vegetation, climatic and floral changes at the Cretaceous-Tertiary boundary. Nature 324:148-152.

Wolfe, J.A. and G.R. Upchurch. 1987. North American nonmarine climates and vegetation during the Late Cretaceous. Palaeogeography, Palaeoclimatology, Palaeoecology 61:33-77.

Wood, D.M. 1991. Homology and phylogenetic implications of male genitalia in Diptera. The ground plan. pp. 255-284 *In* Proceedings of the Second International Congress of Dipterology, Bratislava, Czechoslovakia, 367 pp.

Wood, D.M. and A. Borkent. 1982. Description of the female of *Parasimulium crosskeyi* Peterson (Diptera: Simuliidae) and a discussion of the phylogenetic position of the genus. Memoirs of the Entomological Society of Washington 10:193-210.

Wood, D.M. and A. Borkent. 1989. Phylogeny and classification of the Nematocera. Ch. 114, pp. 1333-1370, *In* Manual of Nearctic Diptera. Vol. 3. Agriculture Canada Monograph 32. vi + pp. 1333-1581.

Wood, D.M., P.T. Dang, and R.A. Ellis. 1979. The mosquitoes of Canada, Diptera: Culicidae. The insects and arachnids of Canada. Part 6. Biosystematics Research Institute, Ottawa, Ontario. 390 pp. + Table 3.

Woodley, N.E. 1989. Phylogeny and classification of the "Orthorrhaphous" Brachycera. Ch. 115, pp. 1371-1395, *In* Manual of Nearctic Diptera. Vol. 3. Agriculture Canada Monograph 32. vi + pp. 1333-1581.

Yoshimoto, C.M. 1975. Cretaceous chalcidoid fossils from Canadian amber. Canadian Entomologist 107:499-528.

Zhang, J.-F. 1986. Luanpingitidae - a new fossil insect family. Acta Palaeontologica Sinica 25:49-54.

Zherikhin, V.V. and I.D. Sukatsheva. 1973. Cretaceous insect-bearing "ambers" (retinites) of northern Siberia. Vopr. Paleontol. Nasekomykh. "Nauka", pp. 3-48.

Zimmerman, R.H. and E.C. Turner. 1984. Dispersal and gonotrophic age of *Culicoides variipennis* (Diptera: Ceratopogonidae) at an isolated site in southwestern Virginia, USA. Journal of Medical Entomology 21:527-535.

TABLES

Table 1.

Numbers of specimens examined and percentage of each species as a portion of the total identified Ceratopogonidae fauna in Canadian Upper Cretaceous amber.

Taxon	*# males*	*# females*	*total #*	*% of total*
Leptoconops primaevus	2	8	10	6.29
Protoculicoides depressus	2	1	3	1.89
Atriculicoides globosus	13	31	44	27.67
Atriculicoides sp.	-	3	3	1.89
Adelohelea glabra	1	2	3	1.89
Heleageron arenatus	4	4	8	5.03
Culicoides filipalpis	12	12	24	15.10
Culicoides tyrrelli	2	7	9	5.66
Culicoides canadensis	5	6	11	6.92
Culicoides bullus	1	2	3	1.89
Culicoides agamus	3	-	3	1.89
Culicoides obuncus	5	-	5	3.14
Culicoides annosus	2	-	2	1.25
Minyohelea pumilis	6	1	7	4.40
Palaeobrachypogon aquilonius	-	2	2	1.25
Palaeobrachypogon remmi	1	3	4	2.51
Palaeobrachypogon vetus	3	2	5	3.14
Peronehelea chrimikalydia	4	9	13	8.17
Specimens identified to species	66	93	159	99.98
Specimens not identified to species	45	50	95	
Total number of specimens	111	143	261*	

* including 7 specimens for which the sex was uncertain

Table 2.
Records of ceratopogonidae in Cretaceous amber. Absolute ages are based on Obradovich (1993).

Location	*Age*	*Absolute Age (in million of years)*	*Number of Species*	*Total Number of Specimens*
Canada	Campanian	71.3 ± 0.5 - 83.5 ± 0.5	18	159
Taimyr Pen., Russia	Coniacian-Santonian	83.5 ± 0.5 - 88.7 ± 0.5	9	58
New Jersey, USA	Cenomanian	93.3 ± 0.2 - 98.5 ± 0.5	1	1
France	Cenomanian	93.3 ± 0.2 - 98.5 ± 0.5	4	4
Lebanon	Hauterivian	127 - 130	1?	?
Golling, Austria	Hauterivian (?)	127 - 130	1?	?
Alaska, USA	Cretaceous	65.4 ± 0.1 - 142	1	1

Table 3.
Descriptive statistics for the antennal ratio (antennal flagellomeres 11-13/1-10) of male Ceratopogonidae in Canadian amber.

Species	*N*	*MIN*	*MAX*	*MEAN*	*SD*
Lep. primaevus	-	-	-	-	-
Pro. depressus	2	.53	.55	.54	-
Atr. globosus	11	.45	.66	.50	.056
Ade. glabra	1	.68	.68	.68	-
Hel. arenatus	1	.56	.56	.56	-
Cul. filipalpis	5	.40	.67	.56	-
Cul. tyrrelli	1	.58	.58	.58	-
Cul. canadensis	3	.61	.65	.63	-
Cul. bullus	-	-	-	-	-
Cul. agamus	1	.65	.65	.65	-
Cul. obuncus	5	.58	.67	.63	-
Cul. annosus	2	.66	.70	.68	-
Cul. casei	1	.65	.65	.65	-
Min. pumilis	2	.29	.30	.29	-
Pal. remmi	-	-	-	-	-
Pal. vetus	2	.64	.65	.65	-
Per. chrimikalydia	3	.54	.66	.60	-

Table 4.
Descriptive statistics for the antennal flagellomeres 10/11 of male Ceratopogonidae in Canadian amber.

Species	*N*	*MIN*	*MAX*	*MEAN*	*SD*
Lep. primaevus	-	-	-	-	-
Pro. depressus	2	.72	.83	.77	-
Atr. globosus	10	.84	1.11	.98	.093
Ade. glabra	1	.50	.50	.50	-
Hel. arenatus	2	.54	.62	.58	-
Cul. filipalpis	6	.46	.55	.52	-
Cul. tyrrelli	1	.40	.40	.40	-
Cul. canadensis	3	.40	.44	.42	-
Cul. bullus	1	.54	.54	.54	-
Cul. agamus	2	.39	.50	.45	-
Cul. obuncus	5	.35	.50	.42	-
Cul. annosus	2	.50	.52	.51	-
Cul. casei	1	.60	.60	.60	-
Min. pumilis	3	1.00	1.00	1.00	-
Pal. remmi	-	-	-	-	-
Pal. vetus	2	.31	.53	.42	-
Per. chrimikalydia	3	.44	.50	.47	-

Table 5.
Descriptive statistics for the mouthpart length/lengtth of fifth tarsomere of foreleg of male Ceratopogonidae in Canadian amber.

Species	*N*	*MIN*	*MAX*	*MEAN*	*SD*
Lep. primaevus	-	-	-	-	-
Pro. depressus	2	2.00	2.38	2.19	-
Atr. globosus	8	1.64	3.29	2.40	.56
Ade. glabra	1	2.22	2.22	2.22	-
Hel. arenatus	2	2.33	2.89	2.61	-
Cul. filipalpis	5	2.39	3.85	3.09	-
Cul. tyrrelli	2	2.84	2.92	2.88	-
Cul. canadensis	3	2.25	2.46	2.32	-
Cul. bullus	1	2.56	2.56	2.56	-
Cul. agamus	2	2.46	2.53	2.50	-
Cul. obuncus	5	2.14	2.92	2.63	-
Cul. annosus	1	2.42	2.43	2.43	-
Cul. casei	1	2.23	2.23	2.23	-
Min. pumilis	1	1.00	1.00	1.00	-
Pal. remmi	1	2.33	2.33	2.33	-
Pal. vetus	2	2.00	2.67	2.33	-
Per. chrimikalydia	4	2.14	3.17	2.72	-

Table 6.
Descriptive statistics for palpal segments 3/4 of male Ceratopogonidae in Canadian amber.

Species	*N*	*MIN*	*MAX*	*MEAN*	*SD*
Lep. primaevus	-	-	-	-	-
Pro. depressus	2	1.64	1.70	1.67	-
Atr. globosus	9	1.54	2.75	2.16	.403
Ade. glabra	1	1.20	1.20	1.20	-
Hel. arenatus	3	1.50	1.67	1.56	-
Cul. filipalpis	7	1.38	2.00	1.67	.215
Cul. tyrrelli	2	1.63	1.75	1.69	-
Cul. canadensis	2	1.14	1.29	1.21	-
Cul. bullus	-	-	-	-	-
Cul. agamus	2	1.33	1.61	1.47	-
Cul. obuncus	4	1.22	2.00	1.52	-
Cul. annosus	2	1.22	1.37	1.30	-
Cul. casei	1	1.33	1.33	1.33	-
Min. pumilis	3	1.75	2.33	2.03	-
Pal. remmi	1	1.17	1.17	1.17	-
Pal. vetus	2	1.12	1.33	1.23	-
Per. chrimikalydia	4	1.43	2.00	1.75	-

Table 7.
Descriptive statistics for wing length (in mm) of male Ceratopogonidae in Canadian amber.

Species	*N*	*MIN*	*MAX*	*MEAN*	*SD*
Lep. primaevus	2	.73	.95	.84	-
Pro. depressus	2	1.16	1.19	1.18	-
Atr. globosus	8	.57	1.05	.84	.163
Ade. glabra	1	.44	.44	.44	-
Hel. arenatus	3	.64	.71	.68	-
Cul. filipalpis	8	.90	1.11	1.02	.075
Cul. tyrrelli	1	.62	.62	.62	-
Cul. canadensis	3	.60	.68	.63	-
Cul. bullus	1	.82	.82	.82	-
Cul. agamus	2	.68	.78	.73	-
Cul. obuncus	4	.71	.89	.80	-
Cul. annosus	2	.73	.76	.74	-
Cul. casei	1	.80	.80	.80	-
Min. pumilis	4	.36	.41	.39	-
Pal. remmi	1	.85	.85	.85	-
Pal. vetus	3	.55	.75	.66	-
Per. chrimikalydia	3	.91	.93	.92	-

Table 8.
Descriptive statistics for the costal ratio (length of costa/wing length) of male Ceratopogonidae in Canadian amber.

Species	*N*	*MIN*	*MAX*	*MEAN*	*SD*
Lep. primaevus	-	-	-	-	-
Pro. depressus	2	.80	.80	.80	-
Atr. globosus	6	.54	.76	.67	-
Ade. glabra	1	.54	.54	.54	-
Hel. arenatus	3	.57	.62	.59	-
Cul. filipalpis	8	.56	.57	.56	.007
Cul. tyrrelli	1	.49	.49	.49	-
Cul. canadensis	3	.47	.54	.50	-
Cul. bullus	1	.56	.56	.56	-
Cul. agamus	2	.51	.52	.51	-
Cul. obuncus	4	.47	.61	.54	-
Cul. annosus	2	.49	.53	.51	-
Cul. casei	1	.62	.62	.62	-
Min. pumilis	4	.27	.32	.30	-
Pal. remmi	1	.43	.43	.43	-
Pal. vetus	2	.56	.57	.56	-
Per. chrimikalydia	3	.56	.62	.59	-

Table 9.
Descriptive statistics for the antennal ratio (antennal flagellomeres 9-13/1-8) of female Ceratopogonidae in Canadian amber. Antennal ratio for *Leptoconops primaevus* is antennal flagellomeres 9-12/1-8.

Species	*N*	*MIN*	*MAX*	*MEAN*	*SD*
Lep. primaevus	4	.75	.77	.76	-
Pro. depressus	1	.95	.95	.95	-
Atr. globosus	13	1.22	1.81	1.45	.174
Atr. sp.	3	1.17	1.28	1.23	-
Ade. glabra	1	1.00	1.00	1.00	-
Hel. arenatus	4	.89	.97	.94	-
Cul. filipalpis	7	.75	1.19	.92	.149
Cul. tyrrelli	5	1.08	1.56	1.19	-
Cul. canadensis	3	.94	1.34	1.15	-
Cul. bullus	2	1.07	1.37	1.22	-
Min. pumilis	1	.60	.60	.60	-
Pal. aquilonius	1	.77	.77	.77	-
Pal. remmi	1	1.29	1.29	1.29	-
Pal. vetus	1	1.07	1.07	1.07	-
Per. chrimikalydia	6	.87	1.19	1.02	-

Table 10.
Descriptive statistics for mouthpart length/length of fifth tarsomere of foreleg of female Ceratopogonidae in Canadian amber.

Species	*N*	*MIN*	*MAX*	*MEAN*	*SD*
Lep. primaevus	3	1.78	2.67	2.17	-
Pro. depressus	-	-	-	-	-
Atr. globosus	11	1.83	2.71	2.27	.267
Atr. sp.	2	2.25	3.08	2.66	-
Ade. glabra	1	3.29	3.29	3.29	-
Hel. arenatus	2	1.77	2.00	1.88	-
Cul. filipalpis	9	3.76	7.50	5.53	1.218
Cul. tyrrelli	5	3.56	4.79	3.95	-
Cul. canadensis	2	2.50	3.53	3.01	-
Cul. bullus	2	2.56	3.64	3.10	-
Min. pumilis	-	-	-	-	-
Pal. aquilonius	2	1.50	1.79	1.64	-
Pal. remmi	2	1.36	1.50	1.43	-
Pal. vetus	1	1.71	1.71	1.71	-
Per. chrimikalydia	5	1.20	2.07	1.58	-

Table 11.
Descriptive statistics for palpal segments 3/4 of female Ceratopogonidae in Canadian amber.

Species	*N*	*MIN*	*MAX*	*MEAN*	*SD*
Lep. primaevus	4	1.08	1.42	1.29	-
Pro. depressus	-	-	-	-	-
Atr. globosus	15	1.29	3.25	2.26	.688
Atr. sp.	3	2.00	3.57	2.71	-
Ade. glabra	1	1.50	1.50	1.50	-
Hel. arenatus	1	1.33	1.33	1.33	-
Cul. filipalpis	10	1.69	3.62	2.43	.624
Cul. tyrrelli	6	1.88	2.78	2.40	-
Cul. canadensis	2	1.50	1.92	1.71	-
Cul. bullus	2	2.67	3.00	2.83	-
Min. pumilis	1	5.00	5.00	5.00	-
Pal aquilonius	2	1.50	2.00	1.75	-
Pal. remmi	3	.92	2.00	1.42	-
Pal. vetus	-	-	-	.	-
Per. chrimikalydia	6	1.42	2.00	1.75	-

Table 12.
Descriptive statistics for wing length (in mm) of female Ceratopogonidae in Canadian amber.

Species	*N*	*MIN*	*MAX*	*MEAN*	*SD*
Lep. primaevus	4	.58	.74	.66	-
Pro. depressus	1	1.63	1.63	1.63	-
Atr. globosus	17	.92	1.26	1.08	. 081
Atr. sp.	3	.88	1.11	.96	-
Ade. glabra	1	.50	.50	.50	-
Hel. arenatus	4	.56	.80	.66	-
Cul. filipalpis	8	.76	1.12	.93	.135
Cul. tyrrelli	6	.56	.94	.81	-
Cul. canadensis	6	.77	1.10	.90	-
Cul. bullus	2	.65	.67	.66	-
Min. pumilis	1	.46	.46	.46	-
Pal aquilonius	2	.65	.74	.69	-
Pal. remmi	2	.87	.98	.93	-
Pal. vetus	1	.71	.71	.71	-
Per. chrimikalydia	6	.76	1.18	.97	-

Table 13.
Descriptive statistics for the costal ratio (length of costa/wing length) of female Ceratopogonidae in Canadian amber.

Species	*N*	*MIN*	*MAX*	*MEAN*	*SD*
Lep. primaevus	3	.38	.44	.40	-
Pro. depressus	1	.89	.89	.89	-
Atr. globosus	17	.73	.84	.78	.028
Atr. sp.	3	.65	.72	.67	-
Ade. glabra	1	.73	.73	.73	-
Hel. arenatus	2	.71	.73	.72	-
Cul. filipalpis	7	.51	.66	.59	.065
Cul. tyrrelli	6	.51	.61	.57	-
Cul. canadensis	6	.52	.70	.61	-
Cul. bullus	2	.62	.64	.63	-
Min. pumilis	1	.37	.37	.37	-
Pal aquilonius	2	.55	.66	.60	-
Pal. remmi	2	.56	.64	.60	-
Pal. vetus	1	.51	.51	.51	-
Per. chrimikalydia	6	.56	.68	.59	-

Table 14.
Descriptive statistics for the hind claw length/hind fifth tarsomere of female Ceratopogonidae in Canadian amber.

Species	*N*	*MIN*	*MAX*	*MEAN*	*SD*
Lep. primaevus	8	.33	.47	.40	.045
Pro. depressus	1	.48	.48	.48	-
Atr. globosus	15	.24	.44	.34	.058
Atr. sp.	3	.29	.47	.37	-
Ade. glabra	2	.33	.40	.37	-
Hel. arenatus	4	.36	.42	.39	-
Cul. filipalpis	11	.25	.43	.35	.055
Cul. tyrrelli	5	.29	.43	.36	-
Cul. canadensis	5	.31	.57	.43	-
Cul. bullus	2	.33	.36	.34	-
Min. pumilis	1	.33	.33	.33	-
Pal aquilonius	2	.47	.50	.48	-
Pal. remmi	3	.81	1.05	.95	-
Pal. vetus	2	1.00	1.24	1.12	-
Per. chrimikalydia	7	1.00	2.14	1.50	.387

Table 15.

Relationship between feeding habits of families of Nematocera and mouthpart structure, exclusive of the Ceratopogonidae. Taxa closely related to Diptera are also included. Families are arranged in a phyletic sequence based on Wood and Borkent (1989). FT = finely toothed. CT = coarsely toothed. SBI = sucking blood from invertebrates. SBV = sucking blood from vertebrates. NSB = not sucking blood. More than one entry for a given family is marked by an asterisk and indicates a variety of conditions present within that family. ? = uncertain status.

Taxon	*Mouthparts*					*Feeding Mode*		
	Mandible			*Lacinia*		*SBI*	*SBV*	*NSB*
	Serrate		*Smooth -absent*	*Toothed*	*Smooth -absent*			
	FT	*CT*						
Siphonaptera			+	+			+	
Nannochoristidae			+		+			+
Tipulidae			+		+			+
Nymphomyiidae			+		+			+
Deuterophlebiidae			+		+			+
Blephariceridae		+			+	+		
Axymyiidae			+		+			+
Pachyneuridae			+		+			+
Bibionidae			+		+			+
Mycetophilidae			+		+			+
Sciaridae			+		+			+
Cecidomyiidae			+		+			+
Psychodidae*			+		+			+
Psychodidae*	+			+			+	
Trichoceridae			+		+			+
Perissommatidae			+		+			+
Anisopodidae			+		+			+
Scatopsidae			+		+			+
Synneuridae			+		+			+
Tanyderidae*	+				?	?		
Tanyderidae*			+		+			+
Ptychopteridae			+		+			+
Dixidae			+		+			+
Corethrellidae*			+		+			+
Corethrellidae*	+				+		+	
Chaoboridae			+		+			+
Culicidae*	+			+			+	
Culicidae*			+	+			+	
Culicidae*			+		+			+
Thaumaleidae			+		+			+
Simuliidae*	+			+			+	
Simuliidae*			+		+			+
Chironomidae*		+			+	?		
Chironomidae*			+		+			+

corrected table

Table 16.

Relationship between feeding habits of families of Culicomorpha and genera of Ceratopogonidae and mouthpart structure. FT = Finely toothed. CT = coarsely toothed. SBI = sucking blood from invertebrates. SBV = sucking blood from vertebrates. NSB = not sucking blood. More than one entry for a given genus is marked by an asterisk and indicates a variety of conditions present within that genus. ? = uncertain status.

Taxon	*Mouthparts*										*SBI*	*SBV*	*NSB*
	Labrum		*Mandible*			*Lacinia*		*Hypopharynx*					
	Toothed	*Smooth / spicules*	*Serrate*		*Smooth*	*Toothed*	*Smooth*	*Toothed*	*Fine spicules*	*Smooth*			
			FT	*CT*									
Dixidae		+			+		+		+				+
Corethrellidae*		+	+				+	+				+	
Corethrellidae*		+			+		+		+				+
Chaoboridae		+			+		+		+				+
Culicidae*		+	+			+				+		+	
Culicidae*		+			+	+				+		+	
Thaumaleidae		+			+		+		+				+
Simuliidae*	+[1]		+			+			+[2]			+	
Simuliidae*		+			+		+		+				+
Chironomidae*		+		+			+		+		?		
Chironomidae*		+			+		+		+				+
Leptoconops*		+	+			+			+			+	
Leptoconops*		+	+			+				+		+	
Austroconops		+	+			+		+				+	
Forcipomyia*		+			+		+		+				+
Forcipomyia*		+	+			+				+		+	
Forcipomyia*		+	+				+			+		+	
Forcipomyia*		+	+			bumps				+	+		
Forcipomyia*		+	+			ridges				+	+		
Forcipomyia*		+		+		prongs				+	+		
Forcipomyia*	+		+			+				+	?		
Atrichopogon*		+			+		+			+			+
Atrichopogon*		+		+		+				+	+		
Atrichopogon*		+	+			+			+		?		?
Atrichopogon*		+	+				+		+		?		?
Atrichopogon*		+	+				+			+			+[3]
Atrichopogon*		+	+				+	+			?		?
Dasyhelea		+			+		+			+			+
Culicoides*	+		+			+		+				+	
Culicoides*		+			+		+		+				+
Culicoides*		+		+		+			+		+		
Other Ceratopogoninae*[4]		+		+			+		+		+		
Other Ceratopogoninae*		+		+			+			+	+		
Other Ceratopogoninae*		+			+		+		+				+

1. Thick, hooked teeth restricted to very apex.
2. Spicules with thick bases.
3. Sucking contents of pollen.
4. Female adults of species of *Camptopterohelea* appear to have fine mandibular teeth but were not examined firsthand. Their biting habits are unknown.

Table 17.

Character states of female mouthparts of Canadian Cretaceous Ceratopogonidae. FT = Finely toothed. CT = coarsely toothed. Hypothesized feeding modes: SBI = sucking blood from invertebrates. SBV = sucking blood from vertebrates. NSB = not sucking blood. No character state noted for a given species indicates that the mouthpart was not visible. (+) signifies that the character state is somewhat uncertain (see species descriptions for further discussion).

on	*Mouthparts*										*SBI*	*SBV*	*NSB*
	Labrum		*Mandible*			*Lacinia*		*Hypopharynx*					
	Toothed	*Smooth / spicules*	*Serrate FT*	*Serrate CT*	*Smooth*	*Toothed*	*Smooth*	*Toothed*	*Fine spicules*	*Smooth*			
;oconops primaevus			+			+						+	
oculicoides depressus													
culicoides globosus	+		+			+						+	
culicoides sp.	+					+						+	
ohelea glabra													
ageron arenatus													
coides filipalpis	+		+			+						+	
coides tyrrelli	+		+			+		+				+	
coides canadensis			(+)			+						+	
coides bullus			+			+						+	
ohelea pumilis													
eobrachypogon uilonius				+							+		
eobrachypogon remmi													
eobrachypogon vetus													
nehelea chrimikalydia		+		+			(+)			+	+		

Table 18.

World Catalog of Fossil Ceratopogonidae (Diptera). All taxa arranged alphabetically after consideration of their subfamily. A total of 175 valid species are listed.

Leptoconopinae

Leptoconops Skuse

Leptoconops Skuse, 1889:288. Type species, *Leptoconops stygius* Skuse, by monotypy.

boreus Kalugina 1991:78. Nizhnyaya Agapa River, 40 km below itssource, Yst'-Yenisey, Krasnoyarsk, Taimyr Peninsula,Russia. Upper Cretaceous.
primaevus Borkent, n. sp. Grassy Lake, Alberta. Upper Cretaceous.
succineus Szadziewski, 1988:233. Denmark. Eocene-Lower Oligocene.

Austroconopinae

Austroconops Wirth and Lee

Austroconops Wirth and Lee, 1959:337. Type species, *Austroconops mcmillani* Wirth and Lee, by monotypy.

borkenti Szadziewski and Schlüter, 1992:78. Durtal, France. Upper Cretaceous.

Forcipomyiinae

Atrichopogon Kieffer

Atrichopogon Kieffer, 1906:53 (as subgenus of *Ceratopogon*). Type species, *Ceratopogon exilis* Coquillett, 1902 [=*Ceratopogon levis* Coquillett], designated by Coquillett, 1910:512.

brunnescens Statz, 1944:141. Rott, Siebengebirge, Germany. Oligocene.
eocenicus Szadziewski, 1988:223. Gdansk, Poland. Eocene-Lower Oligocene.

Atriculicoides Remm

Atriculicoides Remm 1976:108. Type species, *Atriculicoides macrophthalmus*, by original designation.

cenomanensis Szadziewski and Schlüter, 1992:74. Durtal, France. Upper Cretaceous.
globosus (Boesel), 1937:47 (*Lasiohelea*). Cedar Lake, Manitoba. Upper Cretaceous. NEW COMBINATION.
cretea (Boesel), 1937:46 (*Lasiohelea*). Cedar Lake, Manitoba. Upper Cretaceous. NEW SYNONYM, NEW COMBINATION.
incompletus Szadziewski and Schlüter, 1992:76. Durtal, France. Upper Cretaceous.
macrophthalmus Remm, 1976:108. Yantardakh area, Taimyr Peninsula, Russia. Upper Cretaceous.
squamiciliatus Remm, 1976:110. Yantardakh area, Taimyr Peninsula, Russia. Upper Cretaceous.

Forcipomyia Meigen

Forcipomyia Meigen, 1818:73,75. Type species, *Tipula bipunctata* Linnaeus, 1767, designated by Westwood, 1840:126.

berendti Szadziewski, 1988:209. Baltic region. Eocene-Lower Oligocene.
bifidicola Szadziewski, 1993:649. Bitterfeld, Germany. Miocene.
eobreviflagellata Szadziewski, 1988:220. Poland. Eocene-Lower Oligocene.
eocostata Szadziewski, 1988:202. Poland. Eocene-Lower Oligocene.
eophytoheleana Szadziewski, 1988:215. Baltic region. Eocene-Lower Oligocene.
eotrichoheleana Szadziewski, 1988:213. Baltic region. Eocene-Lower Oligocene.
gedanicola Szadziewski, 1988:200. Poland. Eocene-Lower Oligocene.
henningseni Szadziewski, 1988:211. Denmark. Eocene-Lower Oligocene.
krzeminskii Szadziewski, 1988:217. Poland. Eocene-Lower Oligocene.
kulickae Szadziewski, 1988:219. Poland. Eocene-Lower Oligocene.
lyneborgi Szadziewski, 1988:196. Denmark. Eocene-Lower Oligocene.
miocaenica Szadziewski, 1993:647. Bitterfeld, Germany. Miocene.
piriformis (Meunier), 1904:228 (*Ceratopogon*). Baltic region. Eocene-Lower Oligocene.
pseudomicrohelea Szadziewski, 1988:205. Baltic region. Eocene-Lower Oligocene.
subgedanicola Szadziewski, 1993:645. Bitterfeld, Germany. Miocene.
succinea Szadziewski, 1988:194. Poland. Eocene-Lower Oligocene.
succinicola Szadziewski, 1993:650. Bitterfeld, Germany. Miocene.
tuberculosa Szadziewski, 1993:648. Bitterfeld, Germany. Miocene.
turbinata (Meunier), 1904:227 (*Ceratopogon*). Baltic region.Eocene-Lower Oligocene.
uncula (Meunier), 1904:227 (*Ceratopogon*). Baltic region. Eocene-Lower Oligocene.
unculiformis Szadziewski, 1993:646. Bitterfeld, Germany. Miocene.

Protoculicoides Boesel

Protoculicoides Boesel, 1937:50. Type species, *Protoculicoides depressus*, by original designation.

depressus Boesel, 1937:51. Cedar Lake, Manitoba. Upper Cretaceous.

Dasyheleinae

Dasyhelea Kieffer

Dasyhelea Kieffer, 1911a:5. Type species, *Dasyhelea halophila* Kieffer, by original designation.

antiqua Palmer, 1957:266 (as subspecies of *australis*). Near Yermo, California, Miocene.
browneae Pierce, 1966:87. Calico Mountains, California. Miocene.
dara Pierce, 1966:88. Calico Mountains, California. Miocene.
eodicryptoscenica Szadziewski, 1988:228. Poland. Eocene-Lower Oligocene.
gedanica Szadziewski, 1988:226. Gdansk, Poland. Eocene-Lower Oligocene.
judithae Pierce, 1966:88. Calico Mountains, California. Miocene.
kanakoffi Pierce, 1966:89. Calico Mountains, California. Miocene.
miocaenica Szadziewski, 1993:651. Bitterfeld, Germany. Miocene.
stanislavi Szadziewski, 1988:230. Poland. Eocene-Lower Oligocene.
stenoceras Palmer, 1957:271. Near Yermo, California. Miocene.

Ceratopogoninae

Adelohelea Borkent new genus

Adelohelea Borkent, new genus. Type species, *Adelohelea glabra*, by original designation.

glabra Borkent, n.sp. Cedar Lake, Manitoba. Upper Cretaceous.

Alluaudomyia Kieffer

Alluaudomyia Kieffer, 1913:12. Type species, *Alluaudomyia imparunguis* Kieffer, by original designation.

succinea Szadziewski, 1988:112. Poland. Eocene-Lower Oligocene.

Bezzia Kieffer

Bezzia Kieffer, 1899:69. Type species, *Ceratopogon ornatus* Meigen, by original designation.

eocenica Szadziewski, 1988:187. Gdansk, Poland. Eocene-Lower Oligocene.
longipennis Statz, 1944:151. Rott, Siebengebirge, Germany. Oligocene.

Brachypogon Kieffer

Brachypogon Kieffer, 1899:69. Type species, *Ceratopogon vitiosus* Winnertz, by original designation.

balticus Szadziewski, 1988:93. Baltic region. Eocene-Lower Oligocene.
eocenicus Szadziewski, 1988:91. Denmark. Eocene-Lower Oligocene.
gedanicus Szadziewski, 1988:95. Gdansk, Poland. Eocene-Lower Oligocene.
henningseni Szadziewski, 1988:89. Denmark. Eocene-Lower Oligocene.
miocaenicus Szadziewski, 1993:610. Bitterfeld, Germany. Miocene.
polonicus Szadziewski, 1988:88. Poland. Eocene-Lower Oligocene.
prominulus (Meunier), 1904:228 (*Ceratopogon*). Baltic region. Eocene-Lower Oligocene.

Ceratoculicoides Wirth and Ratanaworabhan

Ceratoculicoides Wirth and Ratanaworabhan, 1971:170. Type species, *Helea longipennis* Wirth, by original designation.

danicus Szadziewski, 1988:102. Denmark. Eocene-Lower Oligocene.

Ceratopalpomyia Szadziewski

Ceratopalpomyia Szadziewski, 1988:171. Type species *Ceratopalpomyia eocenica*, by original designation.

eocenica Szadziewski, 1988:173. Denmark. Eocene-Lower Oligocene.

Ceratopogon Meigen

Ceratopogon Meigen, 1803:261. Type species, *Tipula barbicornis* Linnaeus [misidentified, = *Ceratopogon communis* Meigen], by monotypy.

bitterfeldi Szadziewski, 1993:613. Bitterfeld, Germany. Miocene.
ceranowiczi Szadziewski, 1988:77. Poland. Eocene-Lower Oligocene.
crypticus Szadziewski, 1988:70. Denmark. Eocene-Lower Oligocene.
eminens Meunier, 1904:229. Baltic region. Eocene-Lower Oligocene.
forcipiformis Meunier, 1904:235. Baltic region. Eocene-Lower Oligocene.
defectus Meunier, 1904:229. Baltic region. Eocene-Lower Oligocene.
gedanicus Szadziewski, 1988:74. Poland. Eocene-Lower Oligocene.
grogani Szadziewski, 1988:64. Gdansk, Poland. Eocene-Lower Oligocene.
hennigi Szadziewski, 1988:60. Poland. Eocene-Lower Oligocene.
kotejai Szadziewski, 1993:616. Bitterfeld, Germany. Miocene.
margaritae Szadziewski, 1988:80. Gdansk, Poland. Eocene-Lower Oligocene.
miocaenicus Szadziewski, 1993:617. Bitterfeld, Germany. Miocene.
piotrowskii Szadziewski, 1988:75. Denmark. Eocene-Lower Oligocene.
remmicolus Szadziewski, 1988:71. Gdansk, Poland. Eocene-Lower Oligocene.
ritzkowskii Szadziewski, 1988:78. Poland. Eocene-Lower Oligocene.
subeminens Szadziewski, 1993:617. Bitterfeld, Germany. Miocene.
succinicolus Szadziewski, 1993:620. Bitterfeld, Germany. Miocene.
tertiaricus Szadziewski, 1988:62. Gdansk, Poland. Eocene-Lower Oligocene.

Culicoides Latreille

Culicoides Latreille, 1809:251. Type species, *Culicoides punctata* Latreille, by monotypy.

abbreviatipennis Statz, 1944:142. Rott, Siebengebirge, Germany. Oligocene.
abdominalis Statz, 1944:145. Rott, Siebengebirge, Germany. Oligocene (preoccupied by *abdominalis* Santos Abreu, 1918:147).
agamus Borkent, n.sp. Cedar Lake, Manitoba. Upper Cretaceous.
annosus Borkent, n.sp. Cedar Lake, Manitoba. Upper Cretaceous.
atratus Statz, 1944:147. Rott, Siebengebirge, Germany. Oligocene.
austerus Statz, 1944:144. Rott, Siebengebirge, Germany. Oligocene.
balticus Szadziewski, 1988:41. Gdansk, Poland. Eocene-Lower Oligocene.
bicolor Statz, 1944:146. Rott, Siebengebirge, Germany. Oligocene.
bullus Borkent, n.sp. Grassy Lake, Alberta. Upper Cretaceous.
canadensis (Boesel), 1937:47 (*Atrichopogon*). Cedar Lake, Manitoba. Upper Cretaceous. NEW COMBINATION.
carri Pierce, 1966:91. Calico Mountains, California. Miocene.
casei Grogan and Szadziewski 1988:809. Sayreville, Middlesex Co., New Jersey, USA. Upper Cretaceous.
ceranowiczi Szadziewski, 1988:45. Gdansk, Poland. Eocene-Lower Oligocene.
dasyheleiformis Szadziewski, 1988:36. Baltic region. Eocene-Lower Oligocene.
elongatulus Statz, 1944:143. Rott, Siebengebirge, Germany. Oligocene.
eoselficus Szadziewski, 1988:43. Gdansk, Poland. Eocene-Lower Oligocene.
filipalpis Remm, 1976:111. Yantardakh area, Taimyr Peninsula, Russia. Upper Cretaceous.
fossilis Pierce, 1966:91. Calico Mountains, California. Miocene.
gedanensis Szadziewski, 1988:46. Poland. Eocene-Lower Oligocene.
gracilior Statz, 1944:143. Rott, Siebengebirge, Germany. Oligocene.
jucundus Statz, 1944:147. Rott, Siebengebirge, Germany. Oligocene.

kaluginae Remm, 1976:110. Yantardakh area, Taimyr Peninsula, Russia. Upper Cretaceous.
laurae Pierce, 1966:91. Calico Mountains, California. Miocene.
liliputanus Statz, 1944:142. Rott, Siebengebirge, Germany. Oligocene.
megacanthus Palmer, 1957:272. Near Yermo, California. Miocene.
miocenea Pierce, 1966:93. Calico Mountains, California. Miocene.
obesus Statz, 1944:145. Rott, Siebengebirge, Germany. Oligocene.
obscuratus Statz, 1944:144. Rott, Siebengebirge, Germany. Oligocene.
obuncus Borkent, n.sp. Cedar Lake, Manitoba. Upper Cretaceous.
prussicus Szadziewski, 1988:48. Baltic region. Eocene-Lower Oligocene.
speciosus (Meunier), 1904:229 (*Ceratopogon*). Baltic region. Eocene-Lower Oligocene.
sphenostylus Remm, 1976:112. Yantardakh area, Taimyr Peninsula, Russia. Upper Cretaceous.
subgedanensis Szadziewski, 1993:608. Bitterfeld, Germany. Miocene.
succineus Remm, 1976:113. Yantardakh area, Taimyr Peninsula, Russia. Upper Cretaceous.
succivarius Szadziewski, 1988:40. Poland. Eocene-Lower Oligocene.
tenuipennis Statz, 1944:147. Rott, Siebengebirge, Germany. Oligocene.
tyrrelli (Boesel), 1937:48 (*Dasyhelea*). Cedar Lake, Manitoba. Upper Cretaceous. NEW COMBINATION.

Eohelea Petrunkevitch

Eohelea Petrunkevitch 1956:675. Type species, *Eohelea stridulans* Petrunkevitch, by monotypy.
Eohelea Petrunkevitch 1957:208. Type species, *Eohelea stridulans* Petrunkevitch, by original designation. (Primary homonym).

fossicola Szadziewski, 1993:624. Bitterfeld, Germany. Miocene.
gedanica Szadziewski, 1988:165. Poland. Eocene-Lower Oligocene.
grogani Szadziewski, 1988:164. Gdansk, Poland. Eocene-Lower Oligocene.
miocaenea Szadziewski, 1993:625. Bitterfeld, Germany. Miocene.
petrunkevitchi Szadziewski, 1984:39. Gdansk, Poland. Eocene-Lower Oligocene.
petrunkevitchi Szadziewski, 1985:127. Gdansk, Poland. Eocene-Lower Oligocene. (Primary homonym).
sakhalinica Szadziewski, 1990a:78. Firsovo, Starodubskoe, Sakhalinskaja Oblast, Russia. Tertiary.
sinuosa (Meunier), 1904:234 (*Ceratopogon*). Baltic region. Eocene-Lower Oligocene.
stridulans Petrunkevitch, 1956:675. Baltic region. Eocene-Lower Oligocene.
stridulans Petrunkevitch, 1957:209. Baltic region. Eocene-Lower Oligocene (primary homonym).

Fossihelea Szadziewski

Fossihelea Szadziewski, 1988:104. Type species, *Fossihelea gracilitarsis*, by original designation.

gracilitarsis (Meunier), 1904:235 (*Ceratopogon*). Baltic region. Eocene-Lower Oligocene.
miocaenica Szadziewski, 1993:627. Bitterfeld, Germany. Miocene.

Gedanohelea Szadziewski

Gedanohelea Szadziewski 1988:166. Type species, *Gedanohelea loewi*, by original designation.

loewi Szadziewski, 1988:168. Poland. Eocene-Lower Oligocene.
succinea Szadziewski, 1988:169. Gdansk, Poland. Eocene-Lower Oligocene.
wirthi Szadziewski, 1988:170. Denmark. Eocene-Lower Oligocene.

Heleageron Borkent new genus

Heleageron Borkent, new genus. Type species, *Heleageron arenatus*, by original designation.

arenatus Borkent, n.sp. Grassy Lake, Alberta. Upper Cretaceous.

Leptohelea Wirth and Blanton

Leptohelea Wirth and Blanton, 1970:12. Type species, *Leptohelea micronyx* Wirth and Blanton, by original designation.

taimyrica (Remm), 1976:115 (*Baeohelea*). Yantardakh area, Taimyr Peninsula, Russia. Upper Cretaceous.

Mantohelea Szadziewski

Mantohelea Szadziewski, 1988:144. Type species, *Ceratopogon lacus* Meunier, by original designation.

gedanica Szadziewski, 1988:146. Poland. Eocene-Lower Oligocene.
laca (Meunier), 1904:232 (*Ceratopogon*). Baltic region. Eocene-Lower Oligocene.

Meunierohelea Szadziewski[1]

Meunierohelea Szadziewski, 1988:148. Type species, *Meunierohelea nielseni* Szadziewski, by original designation.
Chimaerohelea Debenham, 1988:801. Type species, *Chimaerohelea caligula* Debenham, by original designation.

gedanicola Szadziewski, 1988:153. Gdansk, Poland. Eocene-Lower Oligocene.
miocaenica (Szadziewski), 1993:622 (*Chimaerohelea*). Bitterfeld, Germany. Miocene.
nielseni Szadziewski, 1988:150. Vesterhavet, Denmark. Eocene-Lower Oligocene.
wirthi Szadziewski, 1988:157. Denmark.Eocene-Lower Oligocene.

Minyohelea Borkent new genus

Minyohelea Borkent, new genus. Type species, *Minyohelea pumilis*, by original designation.

pumilis Borkent, n.sp. Grassy Lake, Alberta. Upper Cretaceous.

---

[1] Neal L. Evenhuis (in litt.) has shown that *Meunierohelea* is a senior synonym of *Chimaerohelea*. Debenham's paper was actually published on May 19, 1988 while Szadziewski's paper appeared on March 15, 1988

Monohelea Kieffer

Monohelea Kieffer, 1917:294. Type species, *Monohelea hieroglyphica* Kieffer, by original designation.

baltica Szadziewski, 1988:130. Gdansk, Poland. Eocene-Lower Oligocene.
clunipes (Loew), 1850:30 (*Ceratopogon*). Baltic region. Eocene-Lower Oligocene.

Nannohelea Grogan and Wirth

Nannohelea Grogan and Wirth, 1980:374. Type species, *Ceratopogon bourioni* Clastrier, by original designation.

eocenica Szadziewski, 1988:101. Denmark. Eocene-Lower Oligocene.
grogani Szadziewski, 1988:99. Denmark. Eocene-Lower Oligocene.

Neurohelea Kieffer

Neurohelea Kieffer, 1925:112. Type species, *Ceratopogon luteitarsis* Meigen, by monotypy.

cothurnata (Meunier), 1904:231 (*Ceratopogon*). Baltic region. Eocene-Lower Oligocene.
flagellus Meunier, 1904:230 (*Ceratopogon*). Baltic region. Eocene-Lower Oligocene.

Palaeobrachypogon Borkent new genus

Palaeobrachypogon Borkent, new genus. Type species, *Palaeobrachypogon vetus*, by original designation.

aquilonius (Boesel), 1937:49 (*Ceratopogon*). Cedar Lake, Manitoba. Upper Cretaceous. NEW COMBINATION.
frigidus (Remm), 1976:115 (*Ceratopogon*). Yantardakh area, Taimyr Peninsula, Russia. Upper Cretaceous. NEW COMBINATION.
macronyx (Remm), 1976:113 (*Ceratopogon*). Yantardakh area, Taimyr Peninsula, Russia. Upper Cretaceous. NEW COMBINATION.
remmi Borkent, n.sp. Grassy Lake, Alberta. Upper Cretaceous.
vetus Borkent, n.sp. Grassy Lake, Alberta. Upper Cretaceous.

Palpomyia Meigen

Palpomyia Meigen, 1818:82. Type species, *Ceratopogon flavipes* Meigen, by monotypy.
Miopalpomyia Pierce, 1966:95. Type species, *Miopalpomyia shilo* Pierce, by original designation.
Neopalpomyia Pierce, 1966:95. Type species, *Neopalpomyia freyi* Pierce, by original designation.
Parapalpomyia Pierce, 1966:97. Type species, *Parapalpomyia ryshkoffi* Pierce, by original designation.

edwardsi Cockerell, 1921:469. Gurnet Bay, Isle of Wight, England. Oligocene.
erikae Szadziewski, 1993:642. Bitterfeld, Germany. Miocene.
freyi (Pierce), 1966:95 (*Neopalpomyia*). Calico Mountains, California. Miocene.
jantari Szadziewski, 1988:182. Poland. Eocene-Lower Oligocene.
multispinosa (Pierce), 1966:97 (*Neopalpomyia*). Calico Mountains, California. Miocene.

riedeli Szadziewski, 1988:183. Poland. Eocene-Lower Oligocene.
ryshkoffi (Pierce), 1966:98 (*Parapalpomyia*). Calico Mountains, California. Miocene.
shilo (Pierce), 1966:95 (*Miopalpomyia*). Calico Mountains, California. Miocene.
succinea Szadziewski, 1988:185. Denmark. Eocene-Lower Oligocene.
unca Hong, 1981:58. Fushun, Liaoning Province, China. Eocene.

Peronehelea Borkent new genus

Peronehelea Borkent, new genus. Type species, *Peronehelea chrimikalydia*, by original designation.

chrimikalydia Borkent, n.sp. Grassy Lake, Alberta. Upper Cretaceous.

Physohelea Grogan and Wirth

Physohelea Grogan and Wirth, 1979b:53. Type species, *Neurohelea oedidactyla* Ingram and Macfie, by original designation.

obtusa (Meunier), 1904:235 (*Ceratopogon*). Baltic region. Eocene-Lower Oligocene.
cothurnatula (Meunier), 1904:231 (*Ceratopogon*). Baltic region. Eocene-Lower Oligocene.

Serromyia Meigen

Serromyia Meigen, 1818:83. Type species, *Ceratopogon femoratus* Meigen, by monotypy.

alpheus (Heyden), 1870:251 (*Ceratopogon*). Rott, Siebengebirge, Germany. Oligocene.
colorata Statz, 1944:150. Rott, Siebengebirge, Germany. Oligocene.
austera Statz, 1944:150. Rott, Siebengebirge, Germany. Oligocene.
spinofemorata Statz, 1944:151. Rott, Siebengebirge, Germany. Oligocene.
anomalicornis (Loew), 1850:30 (*Ceratopogon*). Baltic region. Eocene-Lower Oligocene.
polonica Szadziewski, 1988:135. Gdansk, Poland. Eocene-Lower Oligocene.
ryszardi Borkent, 1990:207. Gdansk, Poland. Eocene-Lower Oligocene.
sinuosa Borkent, 1990:207. Gdansk, Poland. Eocene-Lower Oligocene.
spinigera (Loew), 1850:30 (*Ceratopogon*). Baltic region. Eocene-Lower Oligocene.
elongatus (Meunier), 1904:231 (*Ceratopogon*). Baltic region. Eocene-Lower Oligocene.
succinea Szadziewski, 1988:136. Gdansk, Poland. Eocene-Lower Oligocene.

Stilobezzia Kieffer

Stilobezzia Kieffer, 1911b:118. Type species, *Stilobezzia festiva*, by original designation.
falcata (Meunier), 1904:233 (*Ceratopogon*). Baltic region. Eocene-Lower Oligocene.
spinosus (Meunier), 1904:234 (*Ceratopogon*). Baltic region. Eocene-Lower Oligocene.

veterana (Meunier), 1920:897 (*Tetragoneura*). Rott, Siebengebirge, Germany. Oligocene.
goetghebueri Statz, 1944:148. Rott, Siebengebirge, Germany. Oligocene.
kutscheri Szadziewski, 1993:637. Bitterfeld, Germany. Miocene.
saxonica Szadziewski, 1993:638. Bitterfeld, Germany. Miocene.
succinea Szadziewski, 1993:639. Bitterfeld, Germany. Miocene.

Wirthohelea Szadziewski

Wirthohelea Szadziewski, 1988:158. Type species, *Wirthohelea trifida* Szadziewski, by original designation.
trifida Szadziewski, 1988:159. Poland. Eocene-Lower Oligocene.

Unplaced species of Ceratopogoninae

Ceratopogon escheri Giebel, 1856:252. Baltic region. Eocene-Lower Oligocene.
Johannsenomyia hotchkissae Pierce, 1966:94. Calico Mountains, California. Miocene.
Johannsenomyia swinhoei Cockerell, 1919:243. Burma. Oligocene.
Neoculicoides jeanneae Pierce, 1966:94. Calico Mountains, California. Miocene.
Paraculicoides rouseae Pierce, 1966:94. Calico Mountains, California. Miocene.

Unplaced genera of Ceratopogoninae

Neoculicoides Pierce, 1966:93. Type species, *Neoculicoides jeanneae* Pierce, by original designation. (Senior homonym of *Neoculicoides* Boorman and Lane 1979:327).
Paraculicoides Pierce, 1966:94. Type species, *Paraculicoides rouseae* Pierce, by original designation.

Nomen nuda (some discussed by Szadziewski, 1988:10):

Ceratopogon eucerus Keilbach, 1982:348.
Ceratopogon heinei Keilbach, 1982:348.
Ceratopogon longicornus Keilbach, 1982:348.
Ceratopogon minimus Keilbach, 1982:348.
Ceratopogon pectinatus Keilbach, 1982:348.
Ceratopogon pilosus Keilbach, 1982:348.
Ceratopogon terminalis Keilbach, 1982:349.
Ceratopogon ungulatus Keilbach, 1982:349.
Ceratopogon ungulinus Keilbach, 1982:349.
Culicoides jucundus Statz, 1941:5.
Serromyia austera Statz, 1941:6.
Serromyia spinofemorata Statz, 1941:6.

Questionable fossils:

Meunier (1912) described the following species from Madagascar copal which is of Holocene age (Schlüter and Gnielinski 1987:19).
Ceratopogon crassipalpis Meunier, 1912:365.
Ceratopogon madagascariensis Meunier, 1912:365.
Ceratopogon tenuipes Meunier, 1912:366.
Ceratopogon viscatus Meunier, 1912:364.

Figs. 1A-L. Structures of *Leptoconops primaevus*; A-D male, E-L female; A, wing (MCZ 9001); B, antenna (CAS 792); C-D, genitalia, C - (CAS 792), D - (MCZ 9001); E, wing (CAS 614); F, antenna (CAS 335); G, foreleg claws (CAS 733); H, midleg claws (CAS 733); I, hindleg claws (CAS 733); J, oblique anterolateral view of head (CAS 335); K, ventral view of female genitalia (CAS 614); L, lateral view of female genitalia (CAS 614).

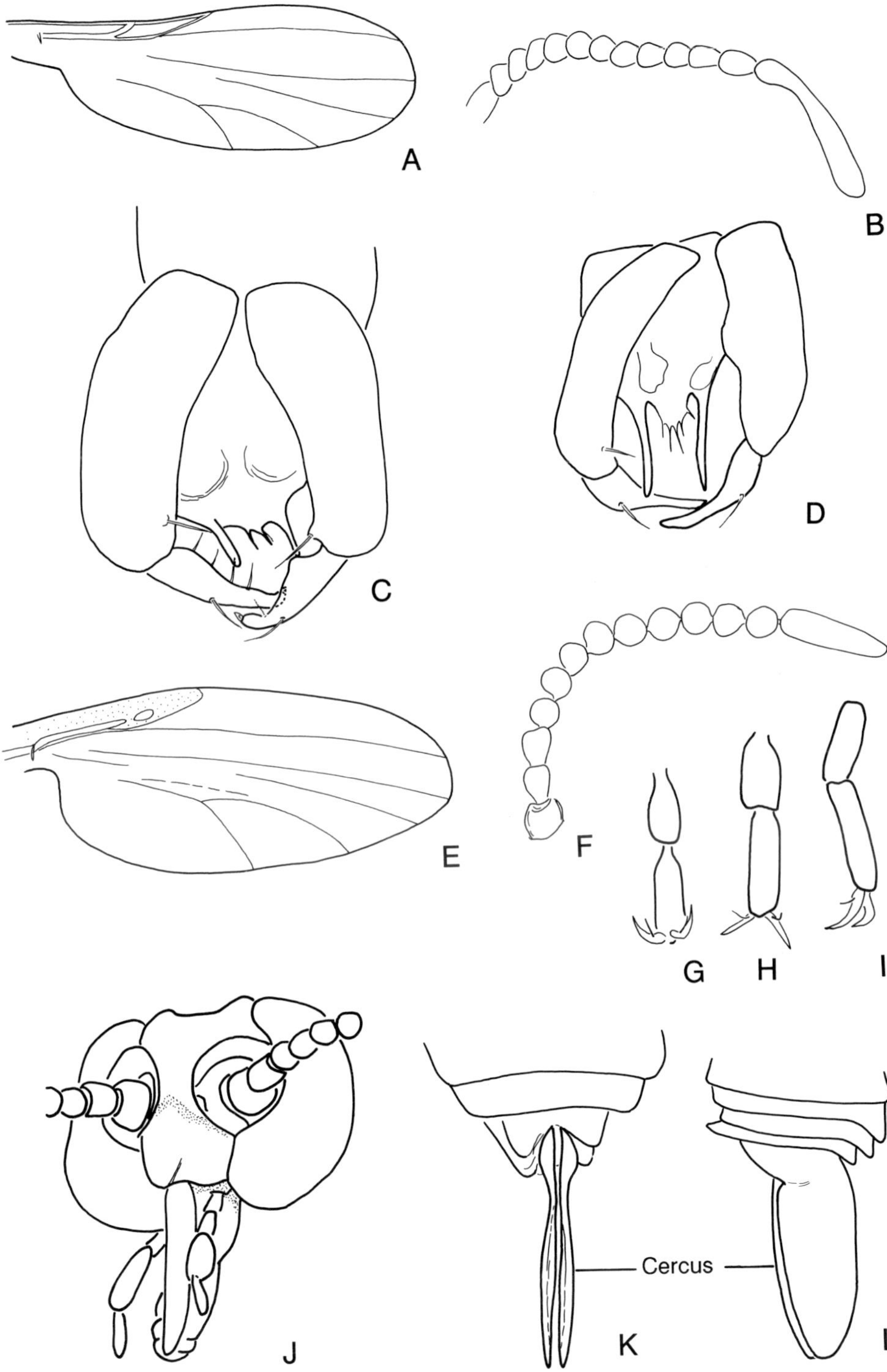
A
B
C
D
E
F
G
H
I
J
Cercus
K
L

Figs. 2A-I. Structures of *Protoculicoides depressus*; A-D male (all CAS 1231), E-I female (all holotype); A, wing; B, palp; C, antenna; D, genitalia; E, wing; F, first tarsomere of hindleg; G, antenna; H, dorsal view of head capsule; I, anterior view of mouthparts.

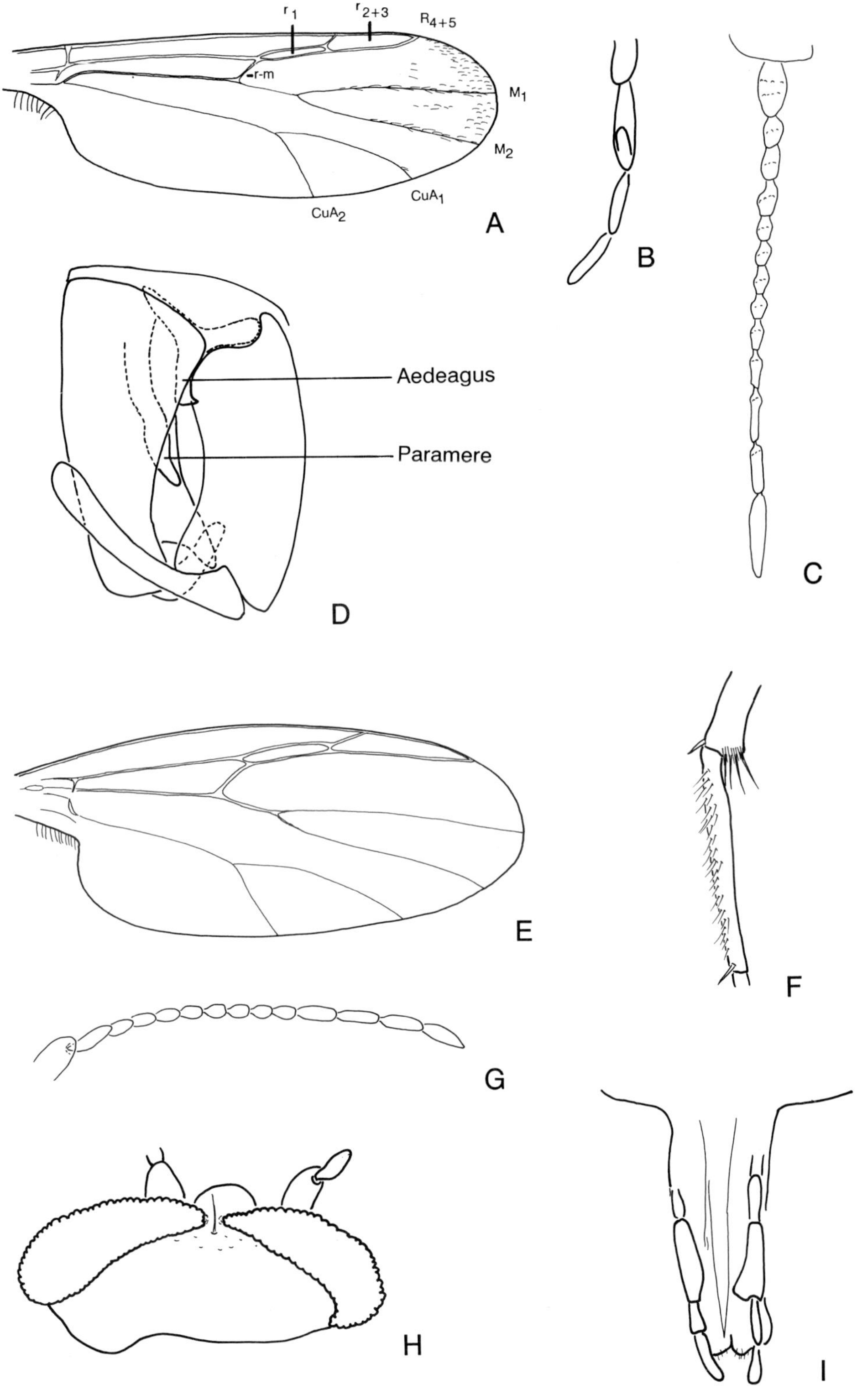
r1
r2+3
R4+5
r-m
M1
M2
CuA1
CuA2
A
B
C
Aedeagus
Paramere
D
E
F
G
H
I

Figs. 3A-H. Structures of male *Atriculicoides globosus*; A, wing (CAS 631); B, antenna (CAS 514); C, first tarsomere of foreleg (ROM 41); D, lateral view of thorax (CAS 1059); E-H, genitalia, E - (CAS 631), F - (CAS 695), G - (MCZ 9003), H - (CAS 532b).

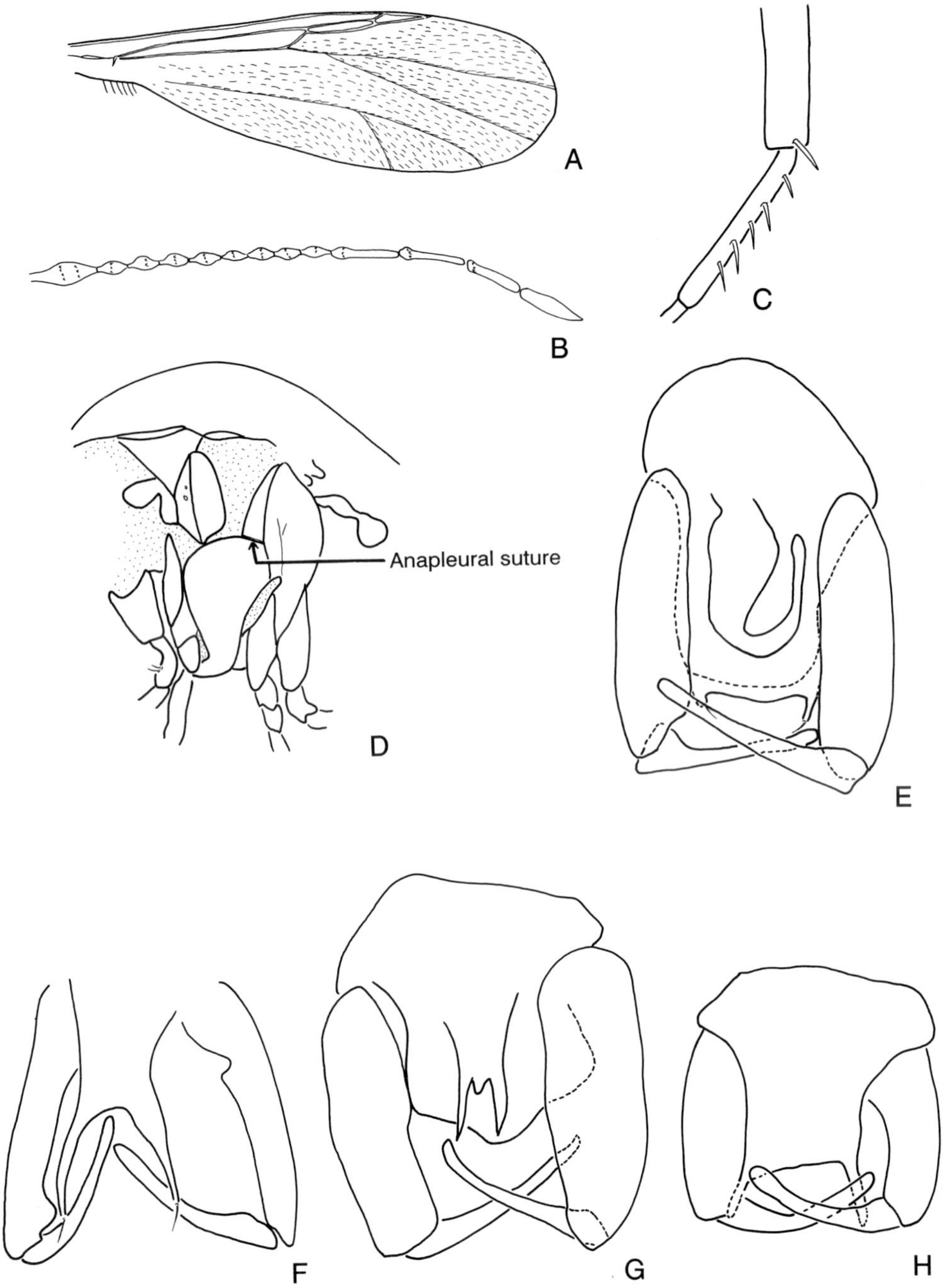
A
B
C
D
Anapleural suture
E
F
G
H

Figs. 4A-M. Structures of *Atriculicoides globosus*; A-B male, C-M female; A-B, genitalia, A - (CAS502), B - (ROM 41); C, wing (holotype); D, antenna (holotype); E, foreleg (holotype); F, midleg (holotype); G, hindleg (CAS 614b); H, anterolateral view of head capsule (MCZ 6894); I, palp (holotype of *A. cretea*); J, mouthparts (CAS 602); K, foreleg claws (holotype); L, midleg claws (holotype); M, hindleg claws (holotype). ? indicates an area of uncertain structure.

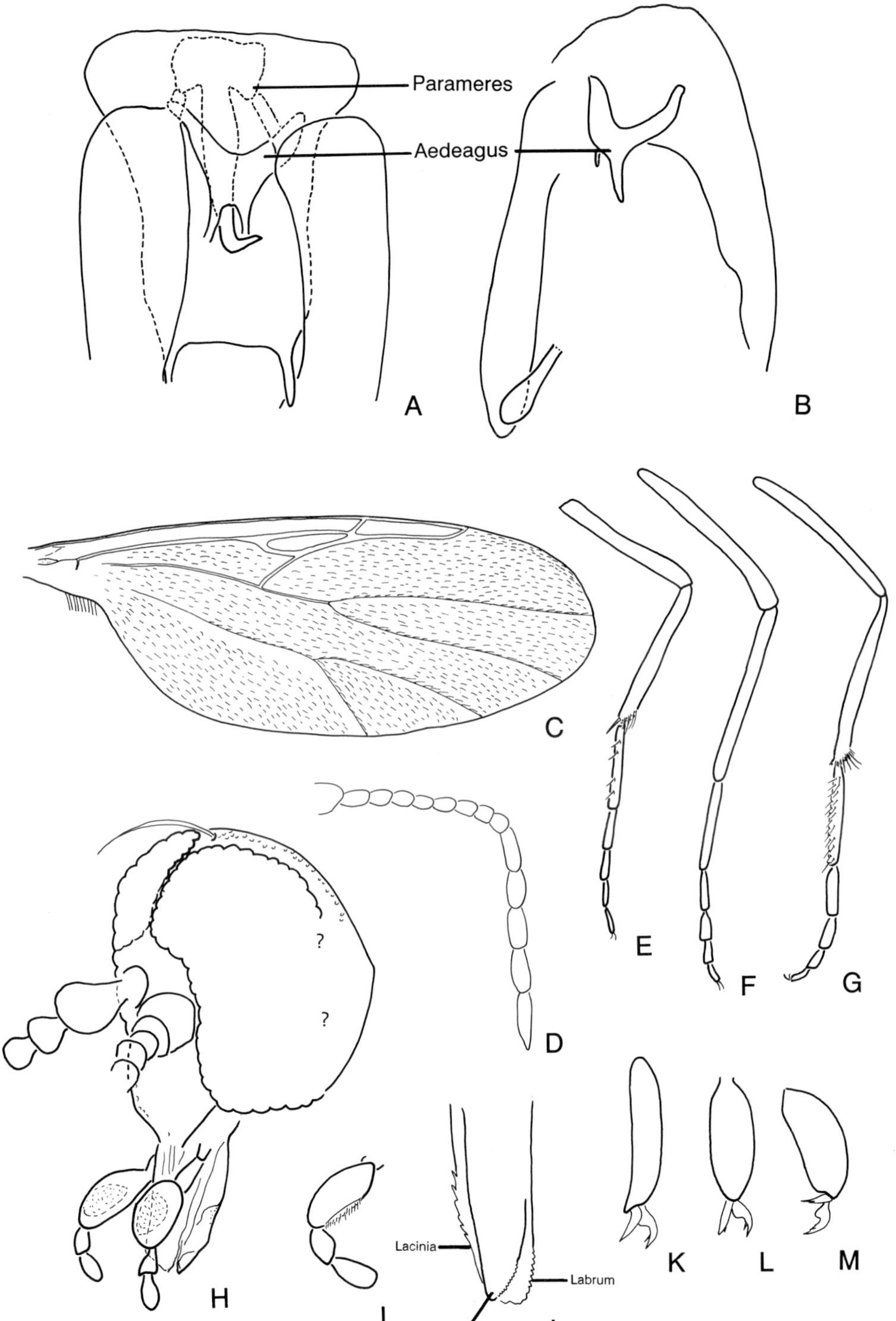
Parameres
Aedeagus
A
B
C
D
E
F
G
H
I
Lacinia
Labrum
Mandible
J
K
L
M

Figs. 5A-O. A-B, genitalia of female *Atriculicoides globosus*; A, ventral view (MCZ 9008); B, ventrolateral view (MCZ 9006); C-O, structures of female *Atriculicoides* sp.; C, wing (ROM 53); D, dorsal view of head capsule (CAS 513); E, antenna (ROM 53); F, ventral view of genitalia (CAS 513); G-H, palp, G - (ROM 53), H - (CAS 513); I, foreleg claws (MCZ 6901); J, midleg claws (MCZ 6901); K, hindleg claws (MCZ 6901); L, hindleg claws (ROM 53); M, foreleg, (CAS 513); N, midleg (CAS 513); O, hindleg (CAS 513). ? indicates an area of uncertain structure.

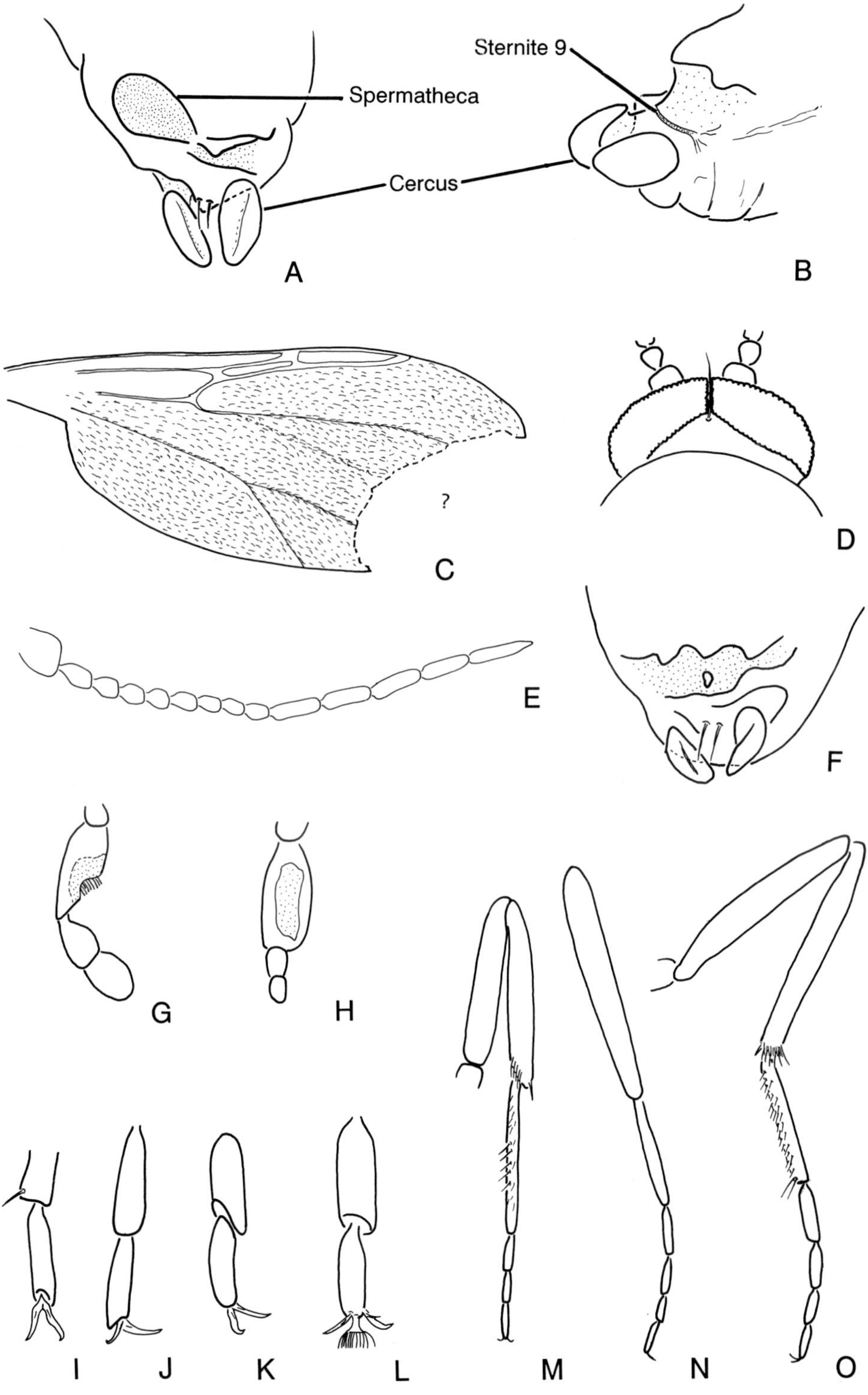
Spermatheca
Sternite 9
Cercus
A
B
C
?
D
E
F
G
H
I
J
K
L
M
N
O

Figs. 6A-Q. A-G, structures of *Adelohelea glabra*; A-C male (holotype), D-G female; A, wing; B, genitalia; C, antenna; D, wing (CAS 1089); E, antenna (P 79-16-12); F, foreleg (CAS 1089); G, hindleg (CAS 1089); H-Q, structures of *Heleageron arenatus*; H-K male, L-Q female; H, genitalia (CAS 985); I-J, antenna, I - (CAS 985), J - (82-15-129); K, tibia and first tarsomere of hindleg (82-15-129); L, basal flagellomeres of antenna (MCZ 9022); M, wing (MCZ 9022); N, antenna (MCZ 9022); O, foreleg claws (MCZ 9022); P, midleg claws (MCZ 9022); Q, hindleg claws (MCZ 9022). s.c.= sensilla coeloconica. ? indicates an area of uncertain structure.

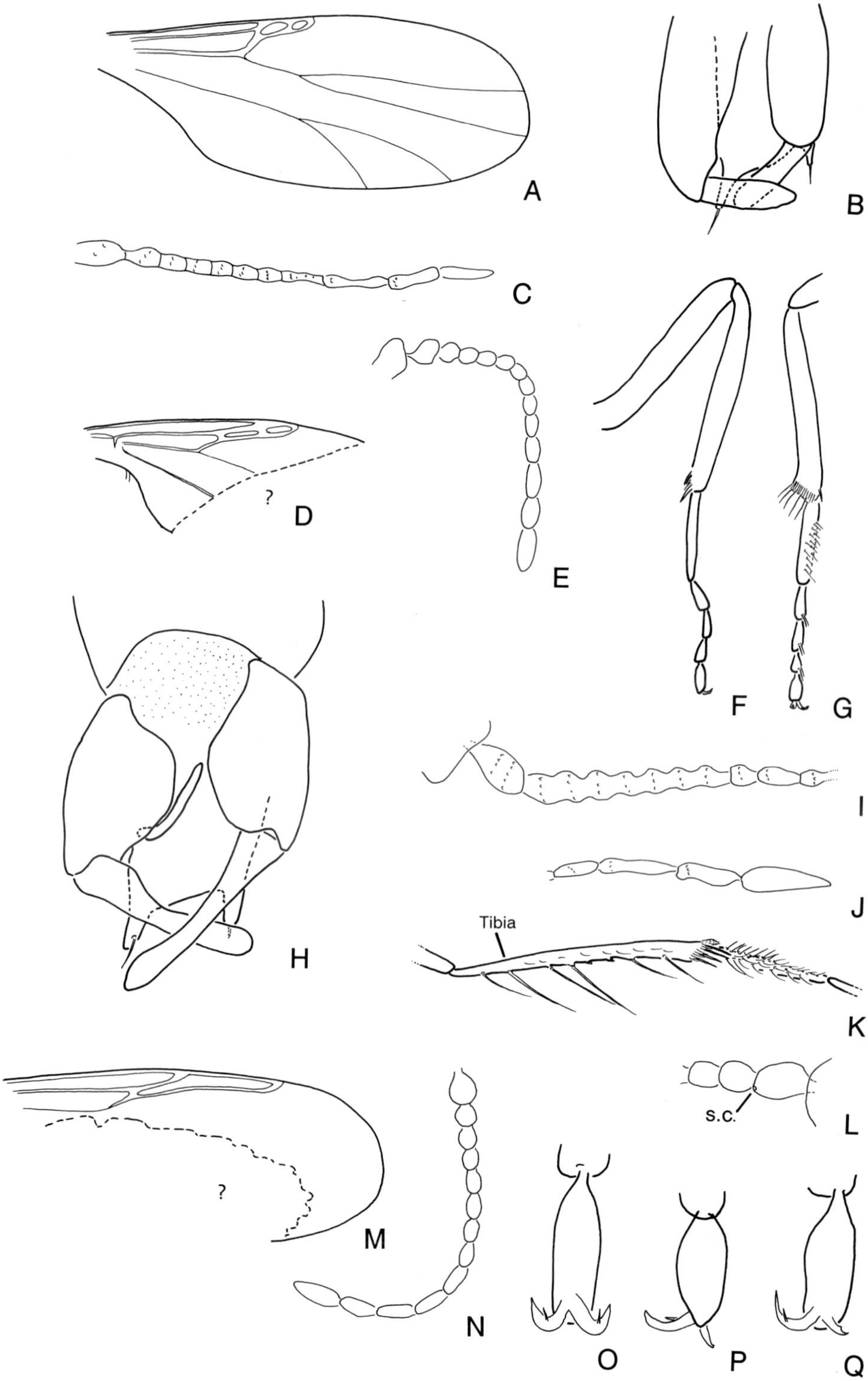
A
B
C
D
?
E
F
G
H
I
J
Tibia
K
s.c.
L
?
M
N
O
P
Q

Figs. 7A-H. Structures of male *Culicoides filipalpis*; A, wing (MCZ 9012); B, antenna (CAS 519); C, palp (CAS 1246); D, genitalia (MCZ 9013); E, anterior view of scutum (CAS 504); F, genitalia (MCZ 9011); G, aedeagus from F; H, parameres from F.

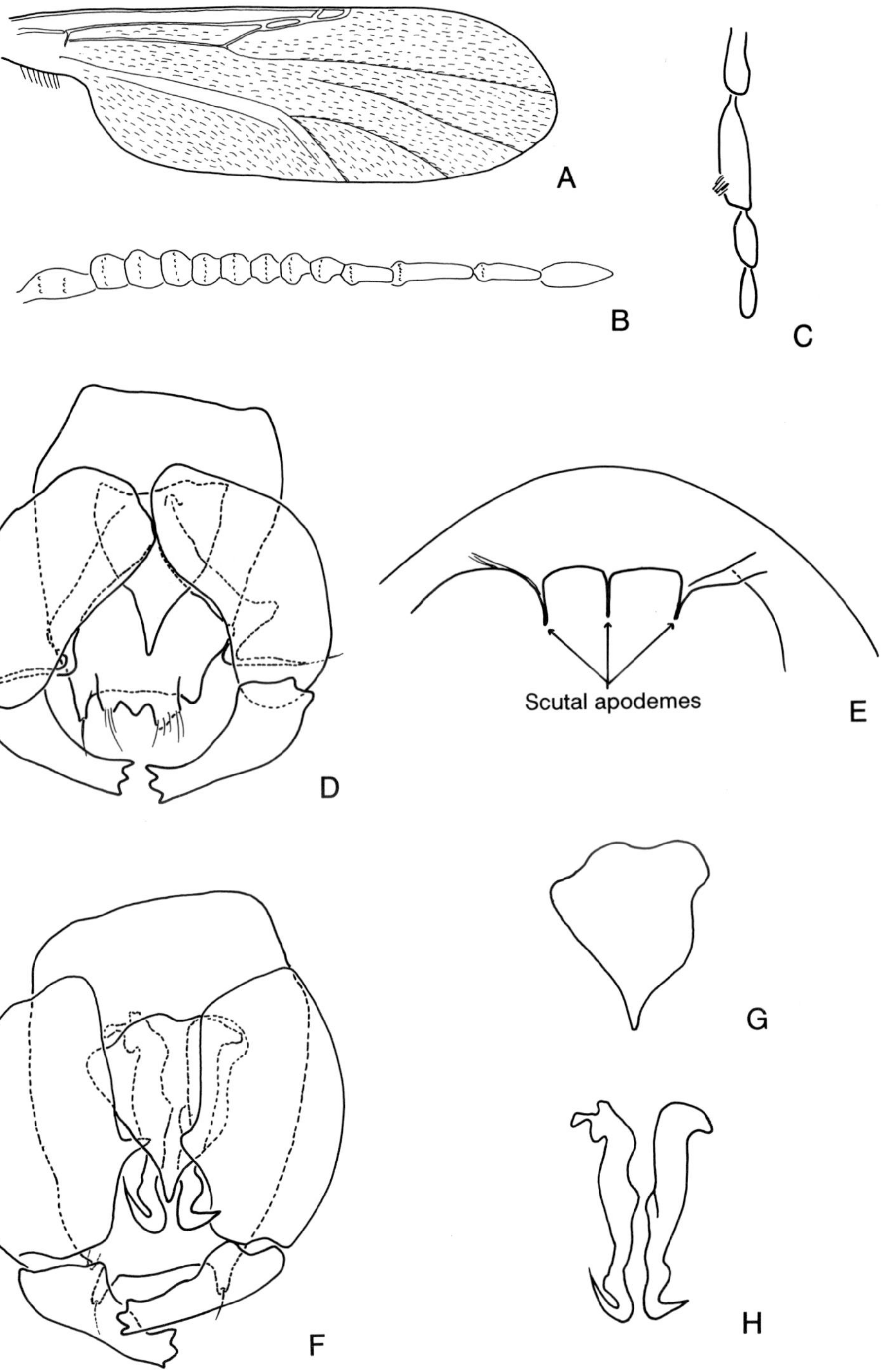
A
B
C
D
Scutal apodemes
E
F
G
H

Figs. 8A-L. Structures of *Culicoides filipalpis*; D male, A-C, E-L female; A, wing (CAS 416); B, antenna (82-15-40); C, foreleg claw (CAS 777); D-E, basal flagellomeres of antenna, D - (MCZ 9013), E - (CAS 667); F, first tarsomere of hindleg (82-15-40); G, lateral view of head capsule (CAS 777); H, mandible (MCZ 9015); I, labrum (MCZ 9015); J, lacinia (82-15-40); K, palp (82-15-40); L, ventral view of genitalia (CAS 416). s.c.= sensilla coeloconica.

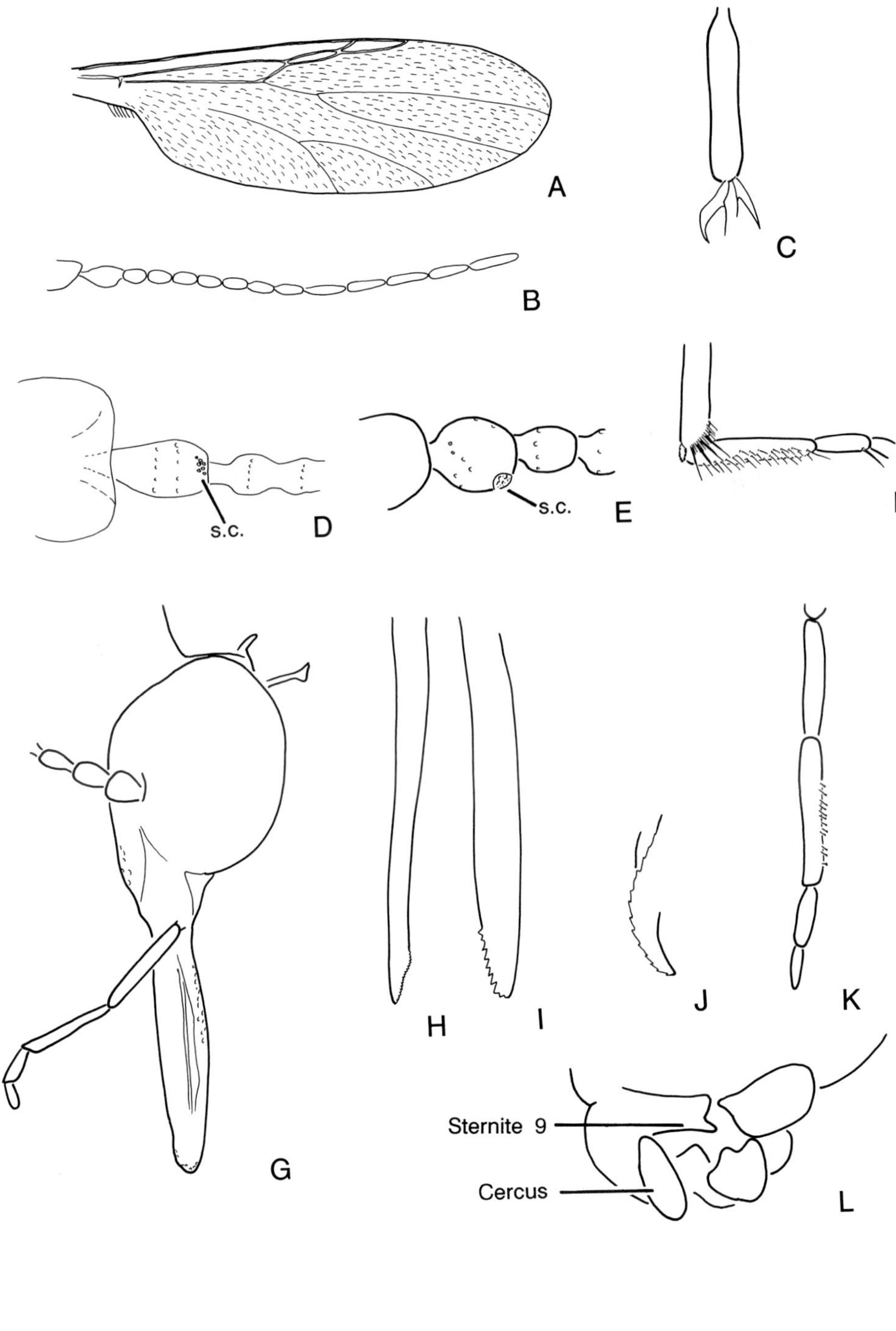
A
B
C
s.c.
D
s.c.
E
F
G
H
I
J
K
Sternite 9
Cercus
L

Figs. 9A-P. Structures of *Culicoides tyrrelli*; A-E male, F-P female; A, wing (CAS 287); B, palp (CAS 287); C, antenna (CAS 287); D-E, genitalia, D - (CAS 287), E - (holotype); F, wing (CAS 944); G, palp (MCZ 9009); H, foreleg (CAS 565); I, midleg claws (CAS 855); J, midleg (CAS 565); K, hindleg (CAS 565); L, lateral view of head capsule (CAS 855); M, antenna (CAS 855); N-O, basal flagellomeres of antenna, N - (CAS 944), O - (ROM 48); P, mouthparts (CAS 944). s.c.= sensilla coeloconica.

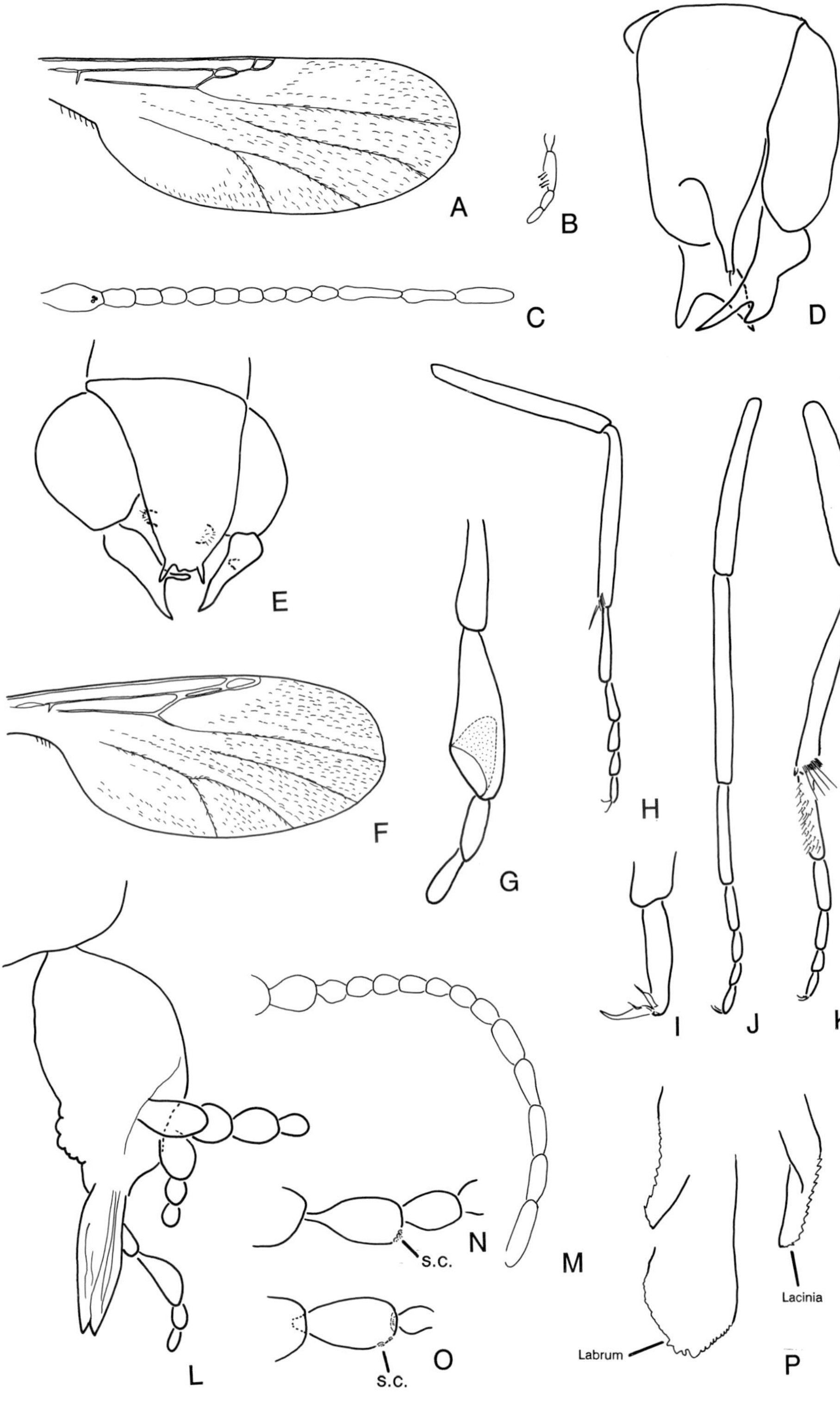

A
B
C
D
E
F
G
H
I
J
K
L
M
N
s.c.
O
s.c.
Lacinia
Labrum
P

Figs. 10A-Q. Structures of *Culicoides canadensis*; A-F male, G-Q female; A, wing (CAS 830); B, antenna (CAS 830); C, basal flagellomeres of antenna (MCZ 9019); D, palp (MCZ 9019); E-F, genitalia, E - (MCZ 9019), F - (CAS 831); G, anterolateral view of head capsule (CAS 1154); H, wing (holotype); I, mouthparts (CAS 721); J, palp (CAS 1154); K, antenna (CAS 507); L, foreleg claws (CAS 516); M, midleg claws (CAS 516); N, hindleg claws (CAS 516); O, midleg (CAS 1154); P, hindleg (CAS 1154); Q, foreleg (CAS 1154). s.c.= sensilla coeloconica.

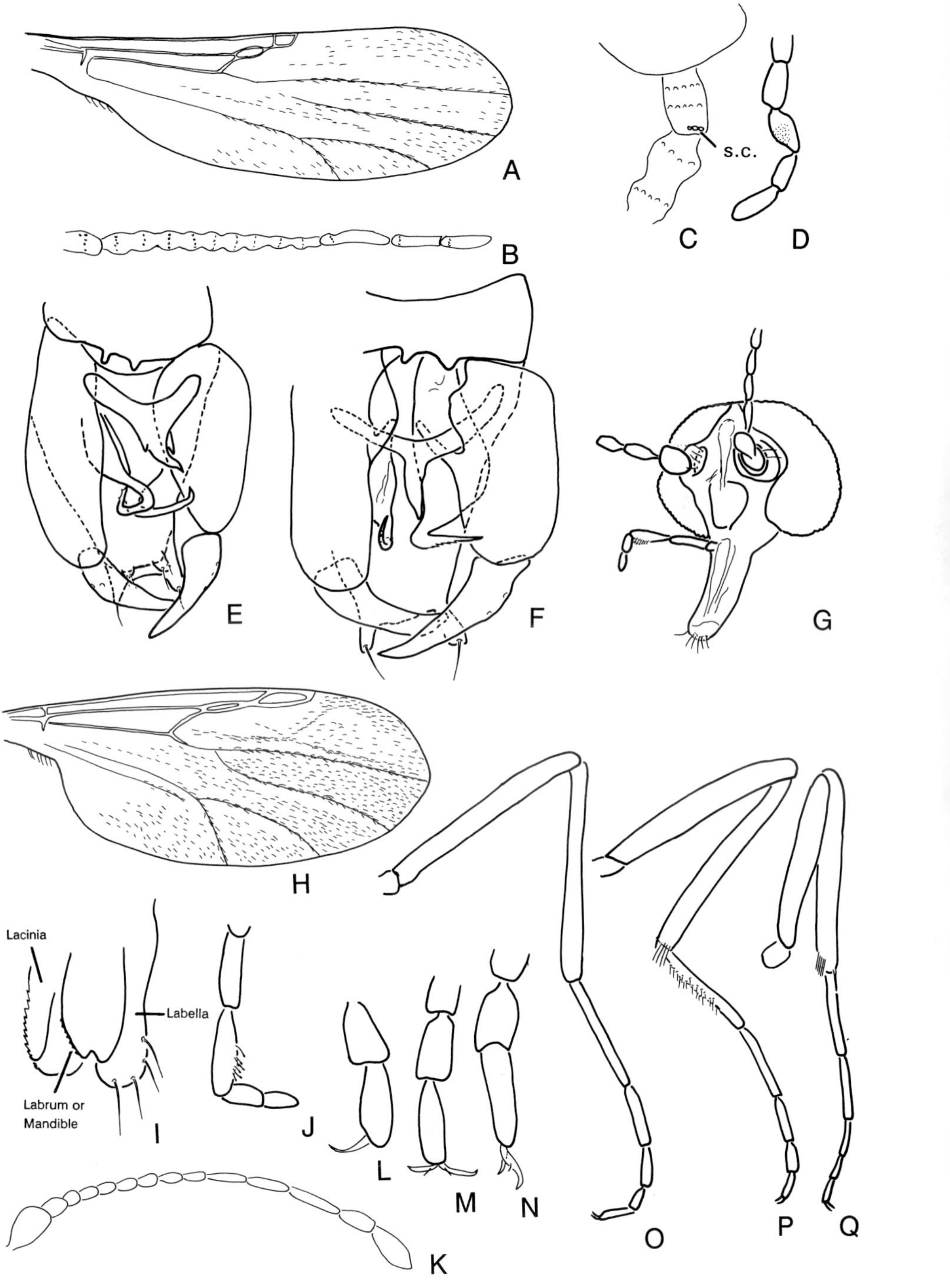
A
s.c.
B
C
D
E
F
G
H
Lacinia
Labella
Labrum or
Mandible
I
J
L
M
N
K
O
P
Q

Figs. 11A-L. A-G, structures of *Culicoides bullus*; A-C male (all holotype), D-G female (all 82-15-20); A, wing; B, antenna; C, genitalia; D, wing; E, palp; F, mandible and lacinia; G, antenna; H-L, structures of male *Culicoides agamus*; H, wing (MCZ 9021); I, antenna (MCZ 9020); J, basal flagellomeres of antenna (MCZ 9020); K, palp (MCZ 9020); L, genitalia (MCZ 9020). s.c.= sensilla coeloconica.

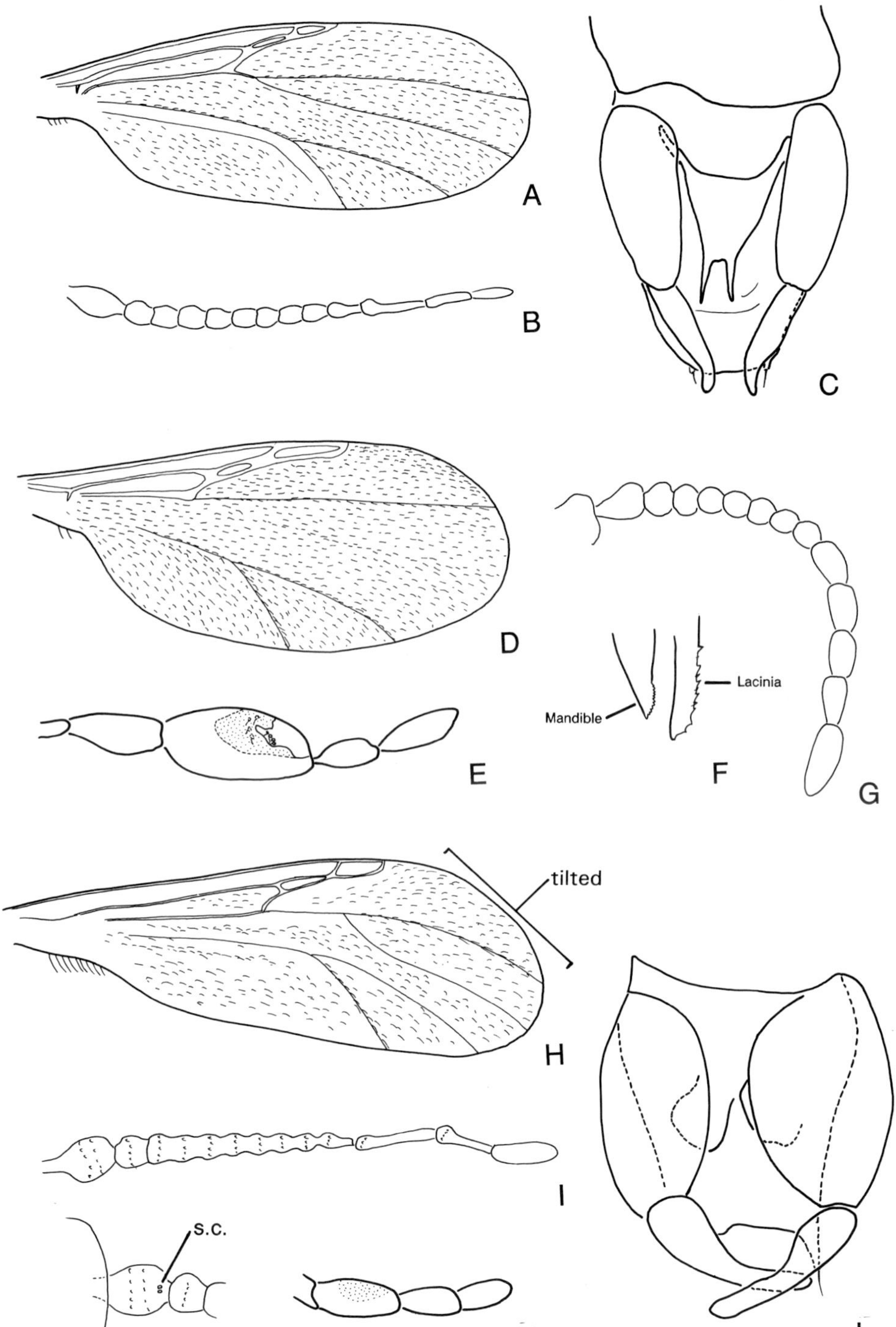
A
B
C
D
E
Mandible
Lacinia
F
G
tilted
H
I
s.c.
J
K
L

Figs. 12A-N. A-H, structures of male *Culicoides obuncus*; A, wing (MCZ 9024); B, antenna (ROM CL-II—59); C, genitalia; D, basal flagellomeres of antenna (CAS 517); E, palp (CAS 1019); F, lateral view of genitalia (CAS 517); G, aedeagus from F; H, parameres from H; I-N, structures of male *Culicoides annosus*; I, wing (holotype); J, antenna (holotype); K, basal flagellomeres of antenna (holotype); L, palp (holotype); M-N, genitalia, M - (CAS 262), N - (holotype). s.c.= sensilla coeloconica. ? indicates an area of uncertain structure.

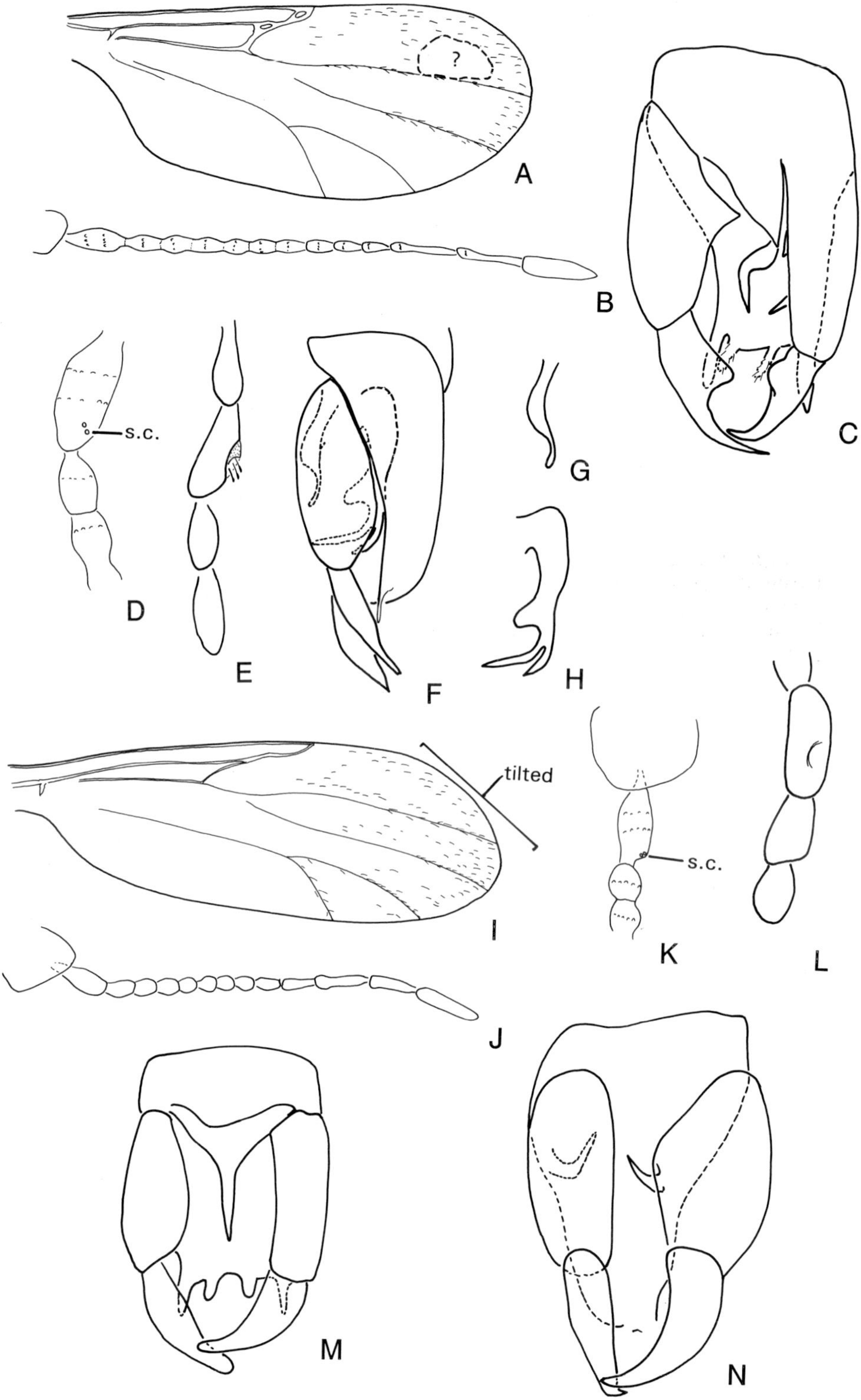
?
A
B
C
s.c.
D
E
F
G
H
tilted
I
J
s.c.
K
L
M
N

Figs. 13A-L. A, male genitalia of *Culicoides casei* (holotype); B-L, structures of *Minyohelea pumilis*; B-E male, F-L female (all CAS 709); B, wing (holotype); C, palp (CAS 131); D, genitalia (CAS 131); E, antenna (holotype); F, wing; G, palp; H, antenna; I, ventral view of genitalia; J, foreleg; K, midleg; L, hindleg.

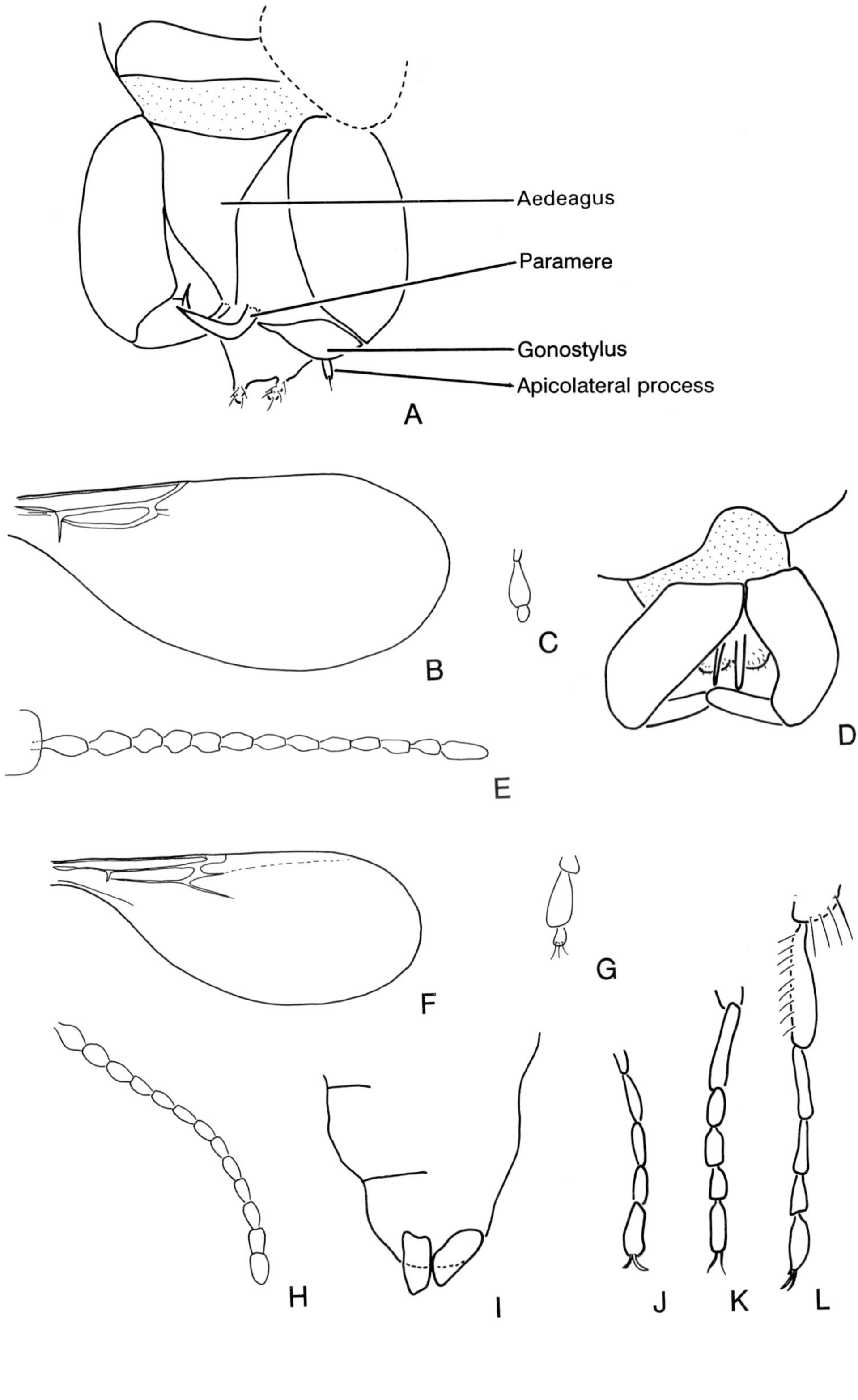
Aedeagus
Paramere
Gonostylus
Apicolateral process
A
B
C
D
E
F
G
H
I
J
K
L

Figs. 14A-O. A-J, structures of female *Palaeobrachypogon aquilonius*; A, wing (82-15-33); B, lateral view of ventral portion of head capsule (holotype); C, foreleg (holotype); D, midleg (holotype); E, hindleg (holotype); F, first tarsomere of hindleg (holotype); G, midleg claws (holotype); H, hindleg claws (holotype); I, antenna (holotype); J, abdomen (holotype); K-O, structures of male *Palaeobrachypogon remmi* (all holotype); K, wing; L, gonostyli from O; parameres from O; N, aedeagus from O; O, genitalia. ? indicates an area of uncertain structure.

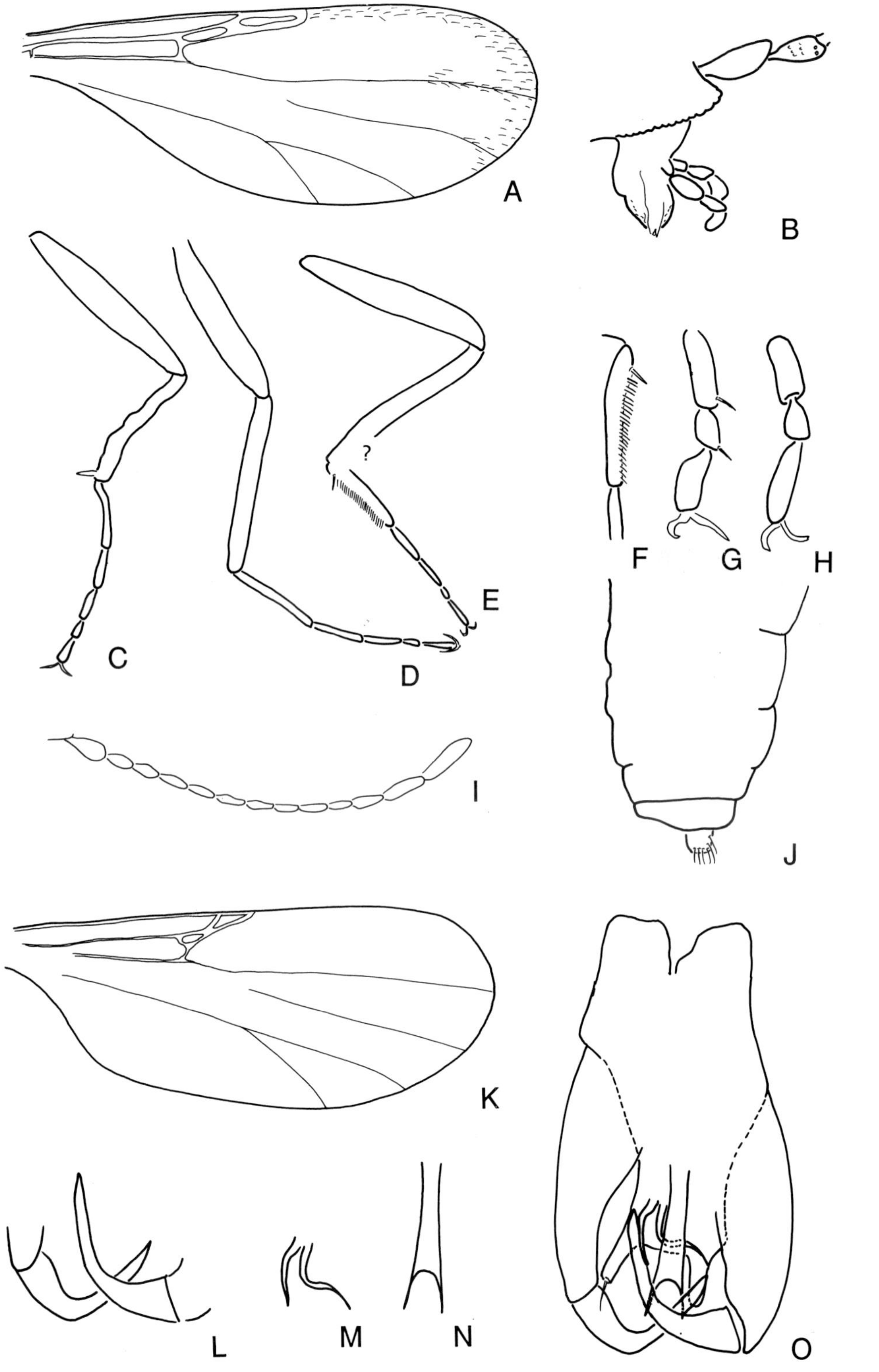

A
B
?
C
D
E
F
G
H
I
J
K
L
M
N
O

Figs. 15A-M. Structures of female *Palaeobrachypogon remmi*; A wing (MCZ 9029); B, lateral view of head capsule (ROM 73); C, palp (MCZ 9030); D, foreleg (MCZ 9029); E, midleg (MCZ 9029); F, hindleg (MCZ 9029); G, antenna (MCZ 9030); H, spermathecae (MCZ 9030); I, hindleg claws (ROM 73); J, foreleg claws (MCZ 9029); K, foreleg claws (MCZ 9029); L, midleg claws (MCZ 9029); M, hindleg claws (MCZ 9029). ? indicates an area of uncertain structure.

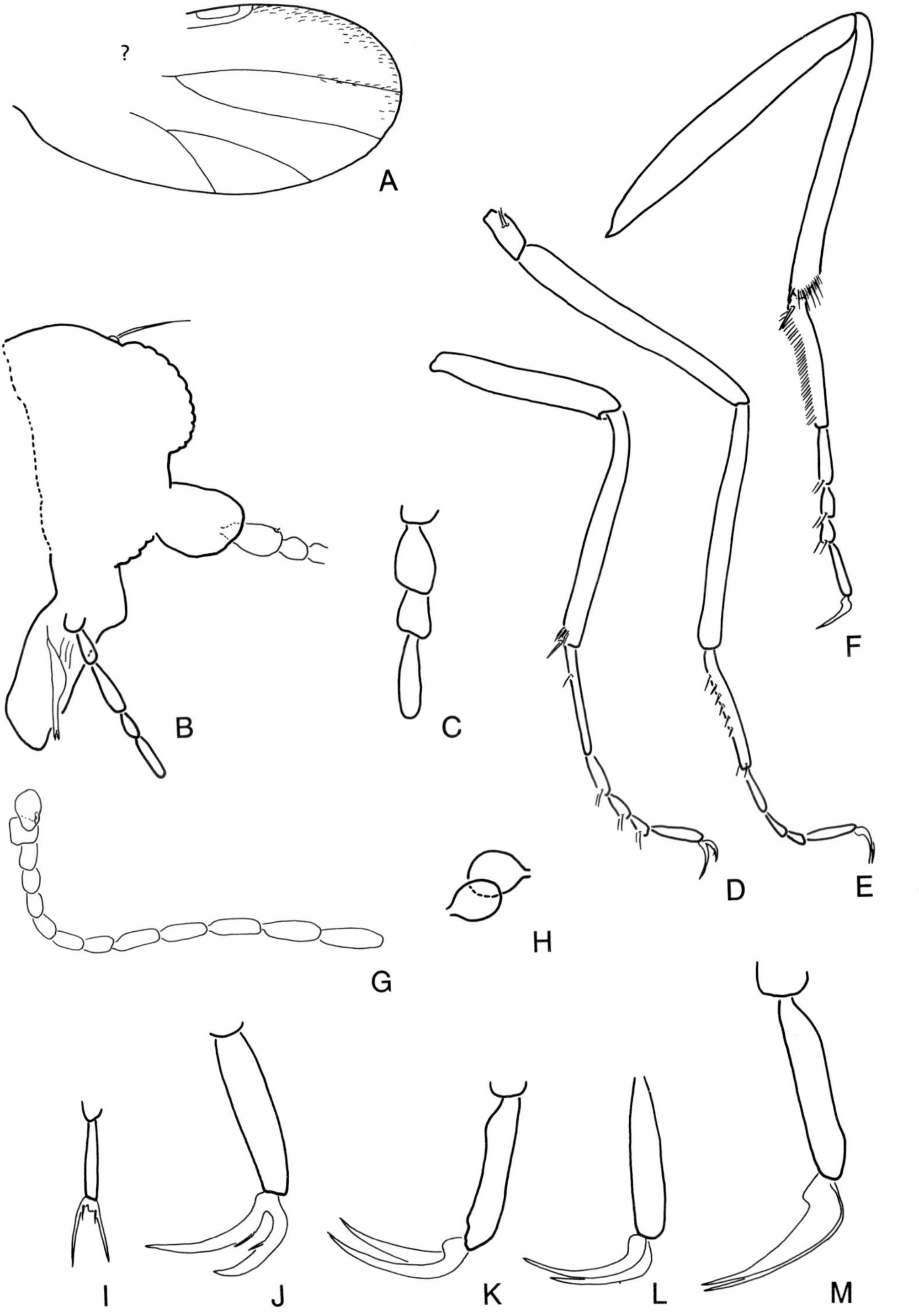

A
?
B
C
D
E
F
G
H
I
J
K
L
M

Figs. 16A-N. Structures of *Palaeobrachypogon vetus*; A-G male, H-N female; A, wing (holotype); B, gonostylus (MCZ 9031); C, parameres (MCZ 9031); D, aedeagus (MCZ 9031); E, parameres (ROM CL-II-58); F, genitalia (holotype); G, antenna (MCZ 9031); H, wing (CAS 692); I, antenna (CAS 692); J, foreleg claws (CAS 764); K, midleg claws (CAS 692); L, hindleg claws (CAS 692); M, hindleg (CAS 692); N, femur and tibia of hindleg (CAS 764). ? indicates an area of uncertain structure.

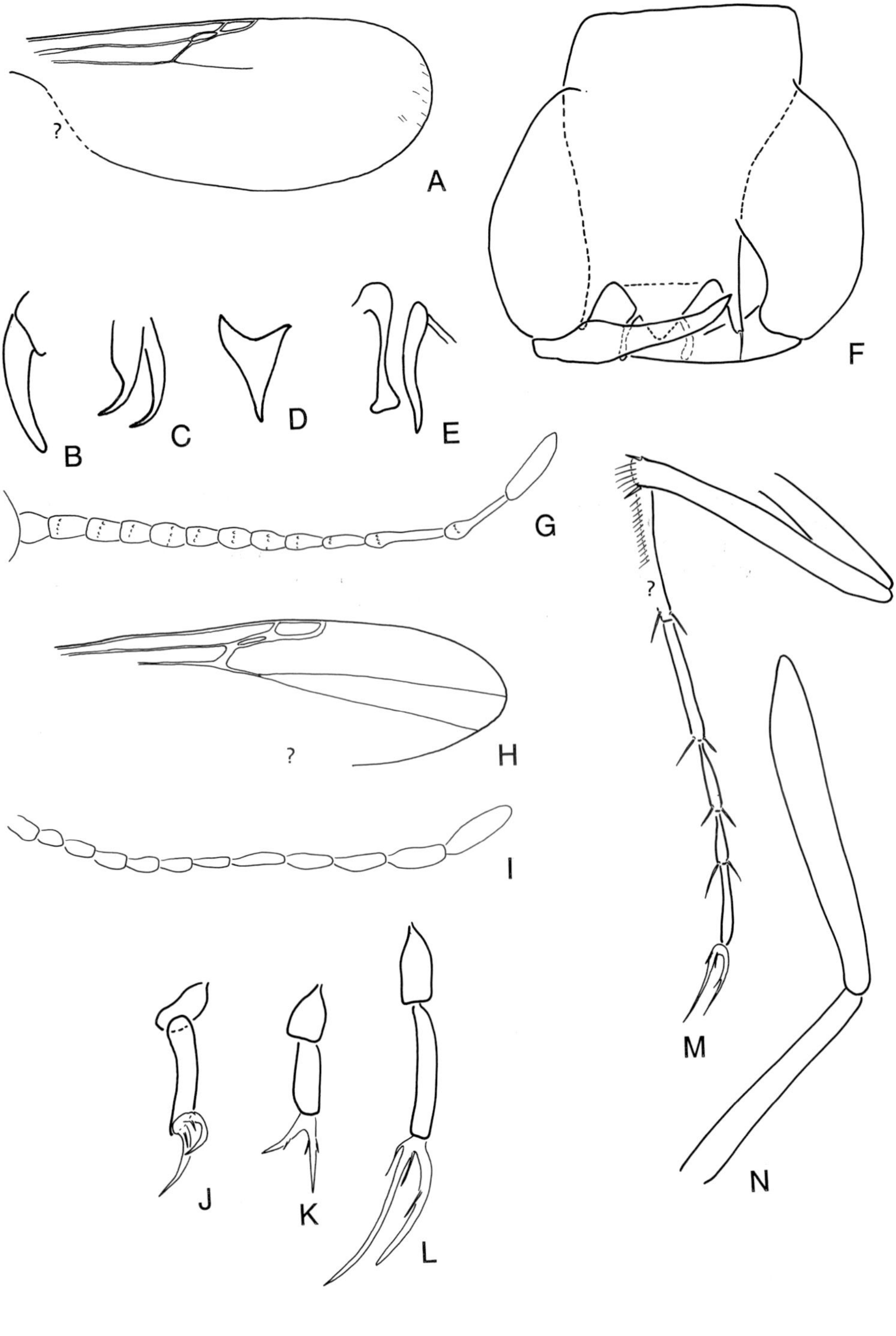
A
B
C
D
E
F
G
H
I
J
K
L
M
N

Figs. 17A-K. Structures of *Peronehelea chrimikalydia*; A-E male, F-K female; A, wing (CAS 339a); B, antenna (MCZ 9025); C-E, genitalia, C - (holotype), D - (CAS 339a), E - (CAS 339a); F, anterolateral view of head capsule (MCZ 9028); G-H, wing, G - (CAS 339b), H - (MCZ 9027); I, antenna (CAS 523); J, palp (MCZ 9028); K, mandibles (MCZ 9028).

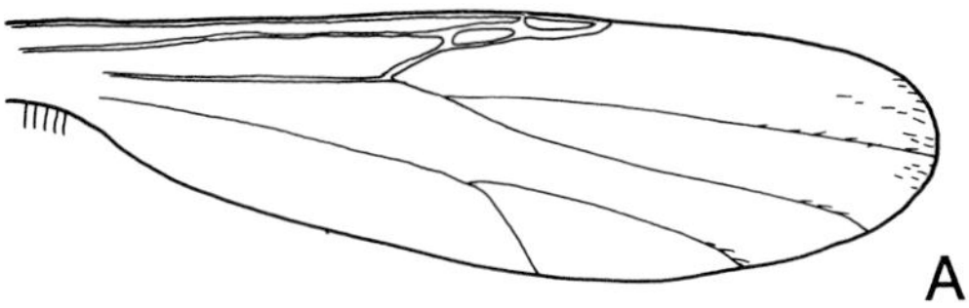
A

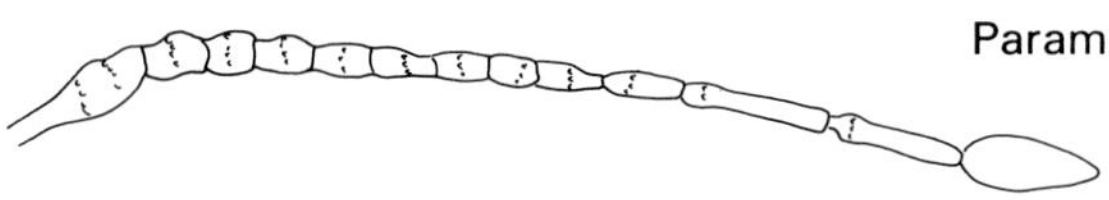
B

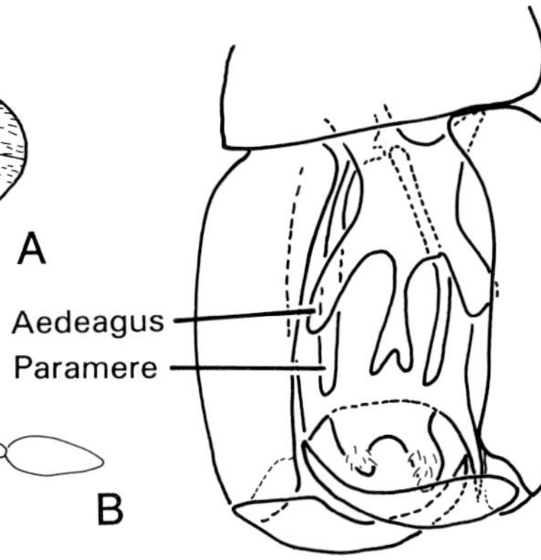

C

D

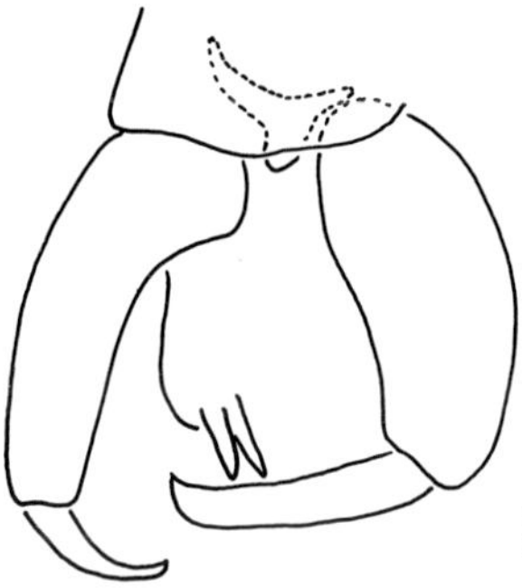
E

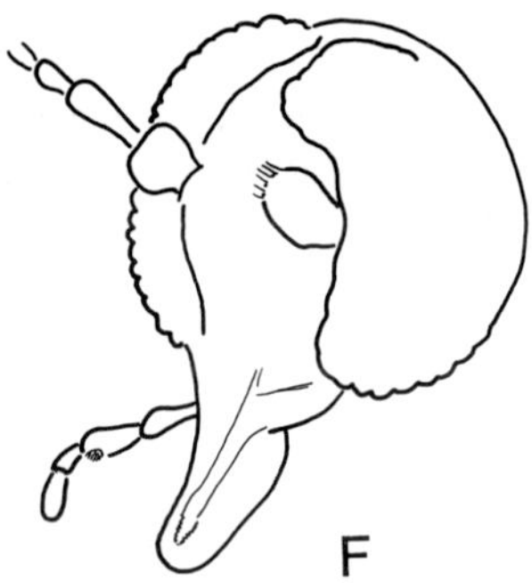
F

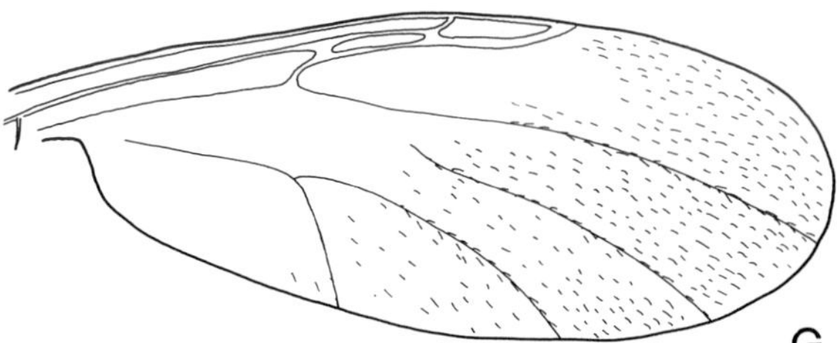
G

J

K

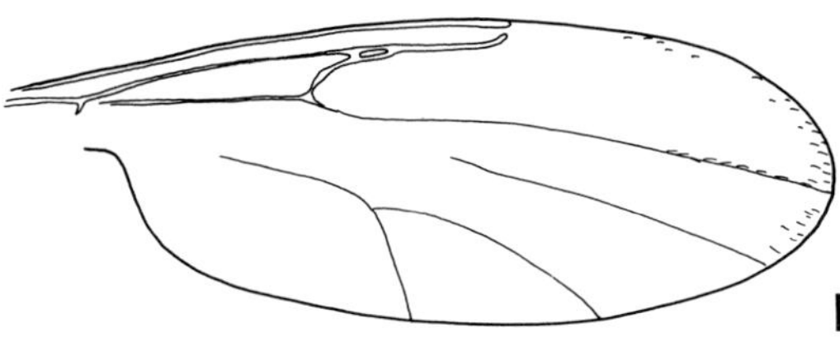
H

I

Figs. 18A-L. A-J, structures of female *Peronehelea chrimikalydia*; A, foreleg (CAS 523); B, midleg (CAS 523); C-D, hindleg, C - (CAS 339b), D - (CAS 523); E, foreleg claws (MCZ 9026); F, midleg claws (MCZ 9026); G-I, hindleg claws, G - (MCZ 9026), H - (MCZ 9028); I, hindleg claws (CAS 339b); J, abdomen with eggs (one protruding posteriorly) (CAS 523); K-L, structures of female *Palaeobrachypogon remmi*; K, lateral view of thorax (MCZ 9029); L, dorsal view of head capsule (MCZ 9030).

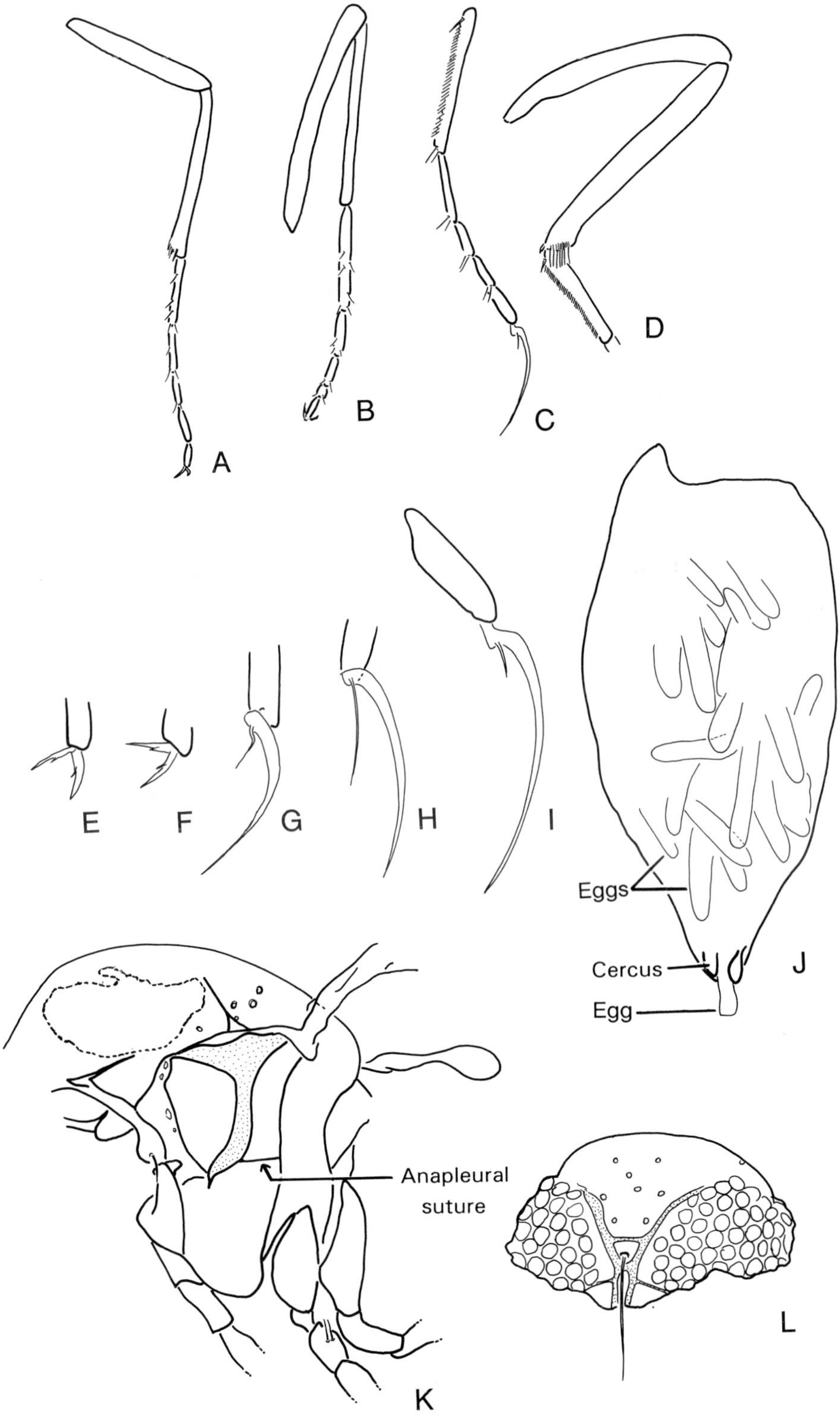
A
B
C
D
E
F
G
H
I
Eggs
Cercus
Egg
J
Anapleural
suture
K
L

Figs. 19A-F. Anterior view of scutum of adult male Ceratopogonidae; A, *Alluaudomyia needhami* Thomsen; B, *Serromyia ledicola* Kieffer; C, *Parabezzia bystraki* Grogan and Wirth; D, *Echinohelea*; E, *Clinohelea*; F, *Palpomyia*.

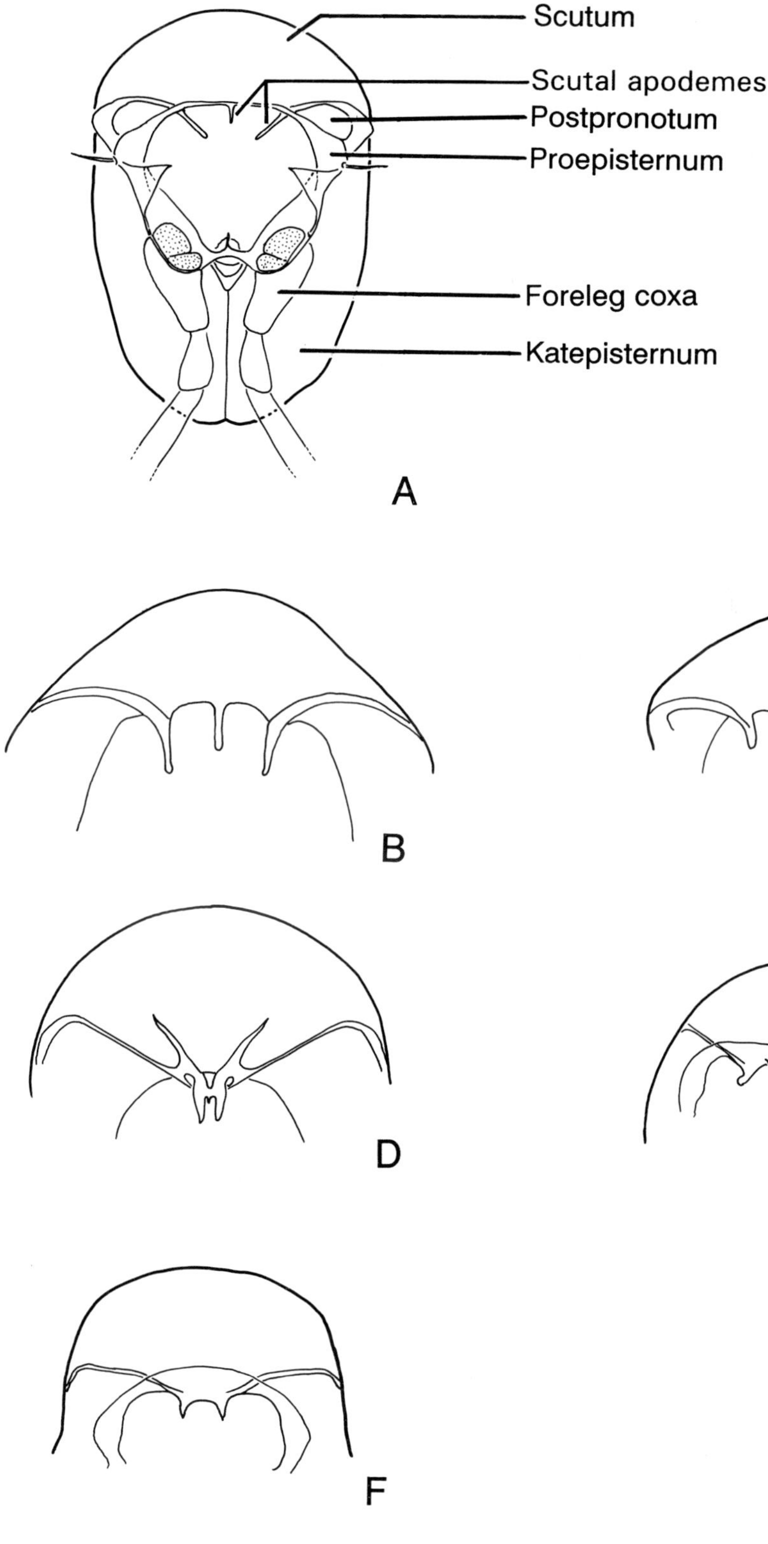
Scutum
Scutal apodemes
Postpronotum
Proepisternum
Foreleg coxa
Katepisternum
A
B
C
D
E
F

Figs. 20A-B. A, North America about 74 million years ago showing location of Grassy Lake, Alberta site (in the west) and Sayreville, New Jersey (in the east). Cedar Lake, Manitoba is not shown as it is clearly an area of secondary deposition. Stippling indicates emergent land. Map compiled from Barron (1987) and map in National Geographic:183(1); B, map showing collection sites of Canadian amber and other western North American localities mentioned in the text.

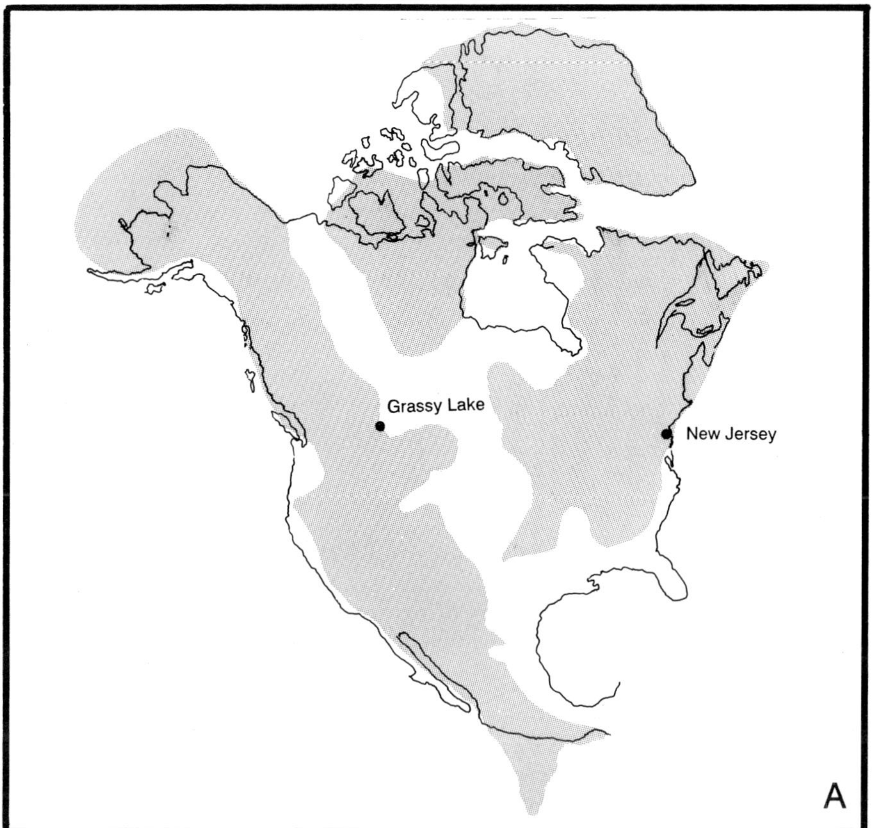
Grassy Lake
New Jersey
A

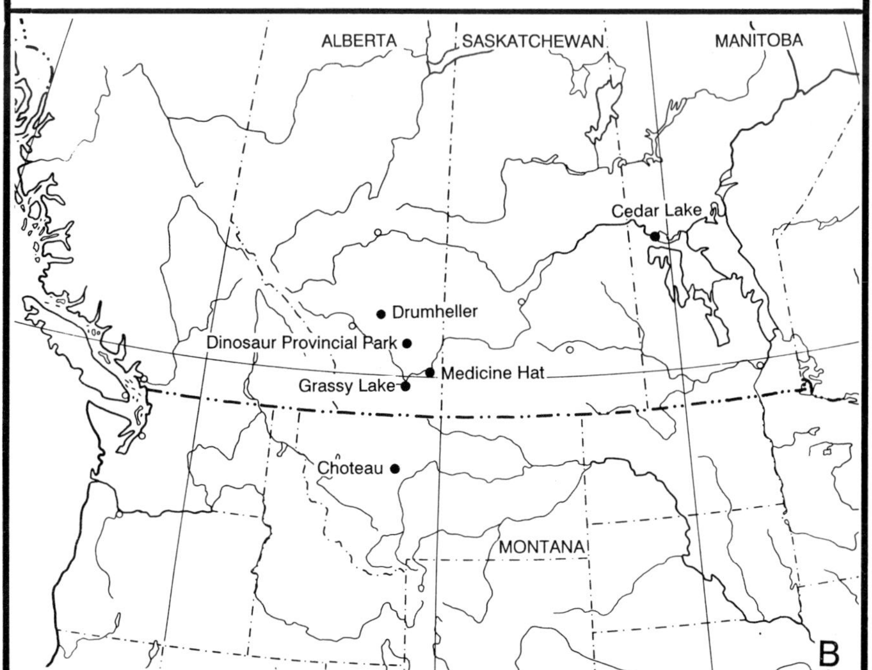
ALBERTA
SASKATCHEWAN
MANITOBA
Cedar Lake
Drumheller
Dinosaur Provincial Park
Medicine Hat
Grassy Lake
Choteau
MONTANA
B

Fig. 21. Cladistic relationships of extant lineages of Ceratopogonidae. Numbers refer to synapomorphies discussed in the text (pp. 87-103).

* Represents unresolved lineages described as the following genera. Each arises as a single lineage from the group delineated by synapomorphies 28" and 42: *Afrohelea, Allohelea, Alluaudomyia, Ankylohelea, Austrohelea, Boreohelea, Borkenthelea, Bothahelea, Bothamia, Cacaohelea, Calcarhelea, Camptopterohelea, Capehelea, Ceratohelea, Chelohelea, Congohelea, Downeshelea, Heterohelea, Kolenohelea, Monohelea, Neohelea, Notiohelea, Notoceratopogon, Paralluaudomyia, Parastilobezzia, Pseudostilobezzia, Schizonyxhelea, Stiloculicoides.*

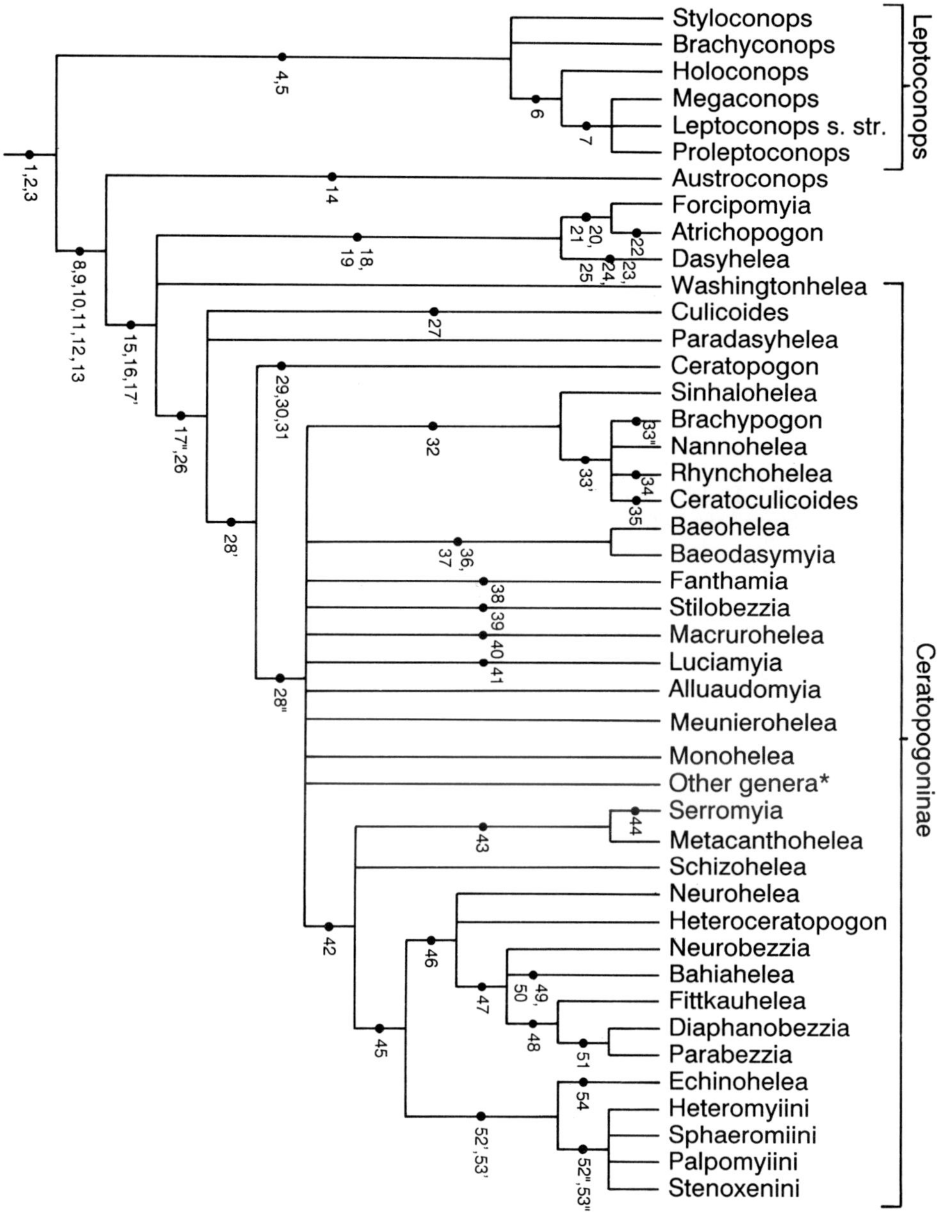
Styloconops
Brachyconops
Holoconops
Megaconops
Leptoconops s. str.
Proleptoconops
Leptoconops
Austroconops
Forcipomyia
Atrichopogon
Dasyhelea
Washingtonhelea
Culicoides
Paradasyhelea
Ceratopogon
Sinhalohelea
Brachypogon
Nannohelea
Rhynchohelea
Ceratoculicoides
Baeohelea
Baeodasymyia
Fanthamia
Stilobezzia
Macrurohelea
Luciamyia
Alluaudomyia
Meunierohelea
Monohelea
Other genera*
Serromyia
Metacanthohelea
Schizohelea
Neurohelea
Heteroceratopogon
Neurobezzia
Bahiahelea
Fittkauhelea
Diaphanobezzia
Parabezzia
Echinohelea
Heteromyiini
Sphaeromiini
Palpomyiini
Stenoxenini
Ceratopogoninae

Figs. 22A-D. Phylogenies of Upper Cretaceous taxa in relation to extant genera. Numbers refer to synapomorphies discussed in the text and presented in Fig. 21. Presented synapomorphies are only those that could be seen in the fossil taxon or those (in the sister groups) which were absent in the fossil. Listed synapomorphies for sister lineages are all those found in that sister group or any lineage within that sister group. A, *Leptoconops*; B, *Protoculicoides*, first hypothesis; C, *Protoculicoides*, second hypothesis; D, *Atriculicoides*.

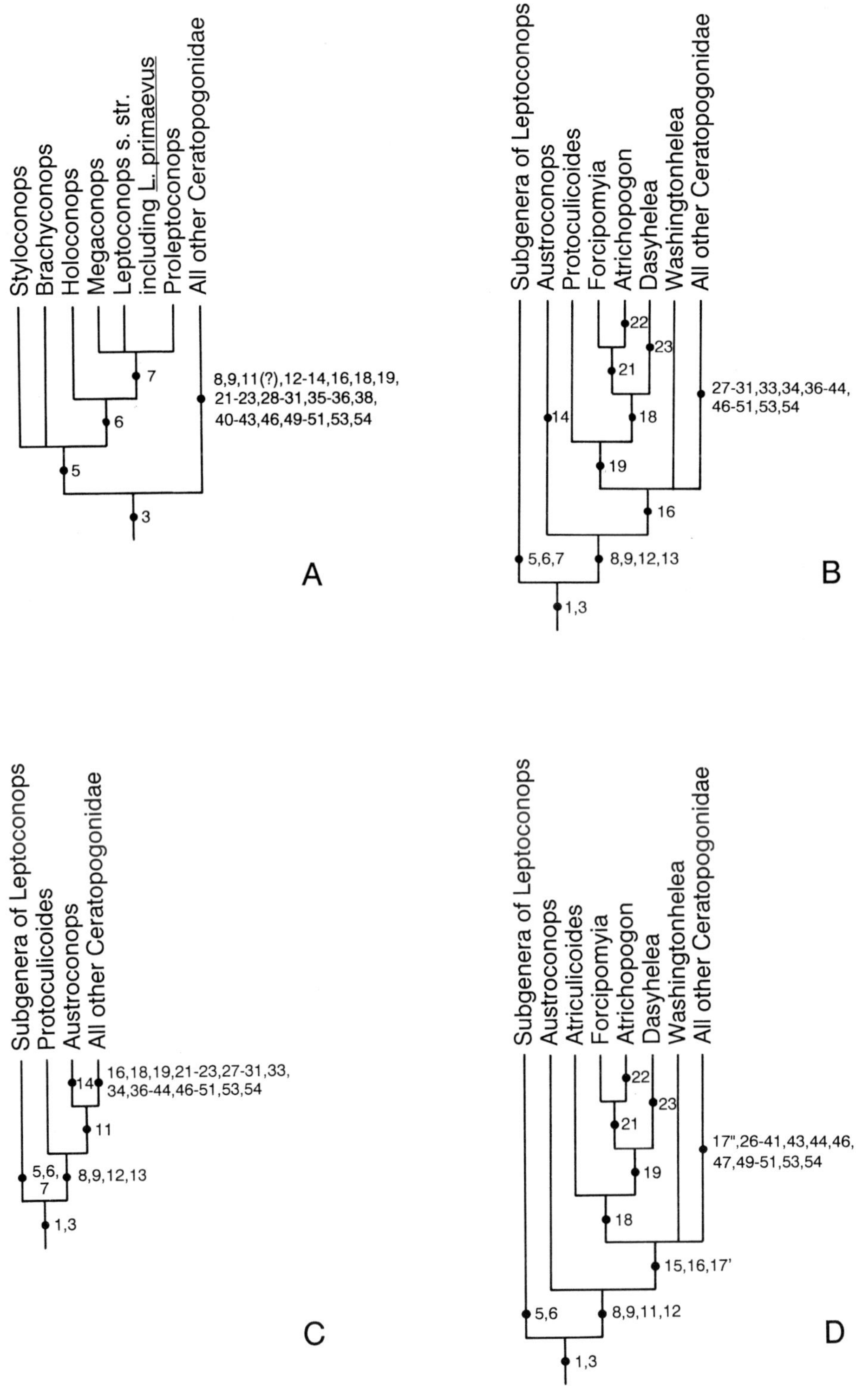

Styloconops
Brachyconops
Holoconops
Megaconops
Leptoconops s. str.
including L. primaevus
Proleptoconops
All other Ceratopogonidae
8,9,11(?),12-14,16,18,19,
21-23,28-31,35-36,38,
40-43,46,49-51,53,54
7
6
5
3
A
Subgenera of Leptoconops
Austroconops
Protoculicoides
Forcipomyia
Atrichopogon
Dasyhelea
Washingtonhelea
All other Ceratopogonidae
22
23
21
27-31,33,34,36-44,
46-51,53,54
14
18
19
16
5,6,7
8,9,12,13
1,3
B
Subgenera of Leptoconops
Protoculicoides
Austroconops
All other Ceratopogonidae
14
16,18,19,21-23,27-31,33,
34,36-44,46-51,53,54
11
5,6,
7
8,9,12,13
1,3
C
Subgenera of Leptoconops
Austroconops
Atriculicoides
Forcipomyia
Atrichopogon
Dasyhelea
Washingtonhelea
All other Ceratopogonidae
22
23
21
17",26-41,43,44,46,
47,49-51,53,54
19
18
15,16,17'
5,6
8,9,11,12
1,3
D

Figs. 23A-D. Phylogenies of Upper Cretaceous taxa in relation to extant genera. Numbers refer to synapomorphies discussed in the text and presented in Fig. 21. Presented synapomorphies are only those that could be seen in the fossil taxon or those (in the sister groups) which were absent in the fossil. Listed synapomorphies for sister lineages are all those found in that sister group or any lineage within that sister group. A, *Adelohelea*; B, *Heleageron*; C, *Culicoides*; D, *Minyohelea*.

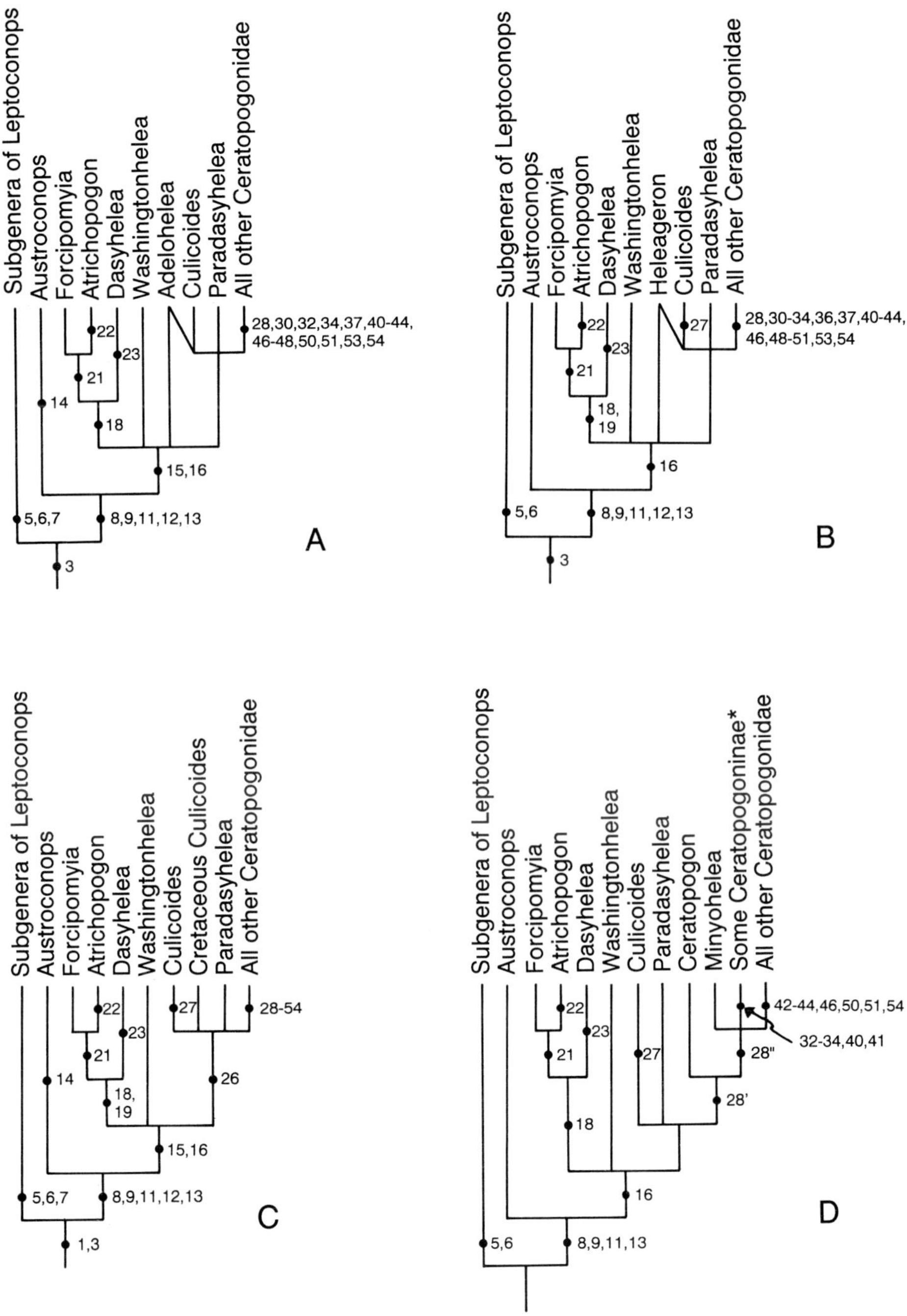
Subgenera of Leptoconops
Austroconops
Forcipomyia
Atrichopogon
Dasyhelea
Washingtonhelea
Adelohelea
Culicoides
Paradasyhelea
All other Ceratopogonidae
28,30,32,34,37,40-44, 46-48,50,51,53,54
22
23
21
14
18
15,16
5,6,7
8,9,11,12,13
3
A
Subgenera of Leptoconops
Austroconops
Forcipomyia
Atrichopogon
Dasyhelea
Washingtonhelea
Heleageron
Culicoides
Paradasyhelea
All other Ceratopogonidae
28,30-34,36,37,40-44, 46,48-51,53,54
22
27
23
21
18, 19
16
5,6
8,9,11,12,13
3
B
Subgenera of Leptoconops
Austroconops
Forcipomyia
Atrichopogon
Dasyhelea
Washingtonhelea
Culicoides
Cretaceous Culicoides
Paradasyhelea
All other Ceratopogonidae
22
27
28-54
23
21
14
26
18, 19
15,16
5,6,7
8,9,11,12,13
1,3
C
Subgenera of Leptoconops
Austroconops
Forcipomyia
Atrichopogon
Dasyhelea
Washingtonhelea
Culicoides
Paradasyhelea
Ceratopogon
Minyohelea
Some Ceratopogoninae*
All other Ceratopogonidae
22
42-44,46,50,51,54
23
32-34,40,41
21
27
28"
28'
18
16
5,6
8,9,11,13
D

Figs. 24A-B. Phylogenies of Upper Cretaceous taxa in relation to extant genera. Numbers refer to synapomorphies discussed in the text and presented in Fig. 21. Presented synapomorphies are only those that could be seen in the fossil taxon or those (in the sister groups) which were absent in the fossil. Listed synapomorphies for sister lineages are all those found in that sister group or any lineage within that sister group. A, *Palaeobrachypogon*; B, *Peronehelea*.

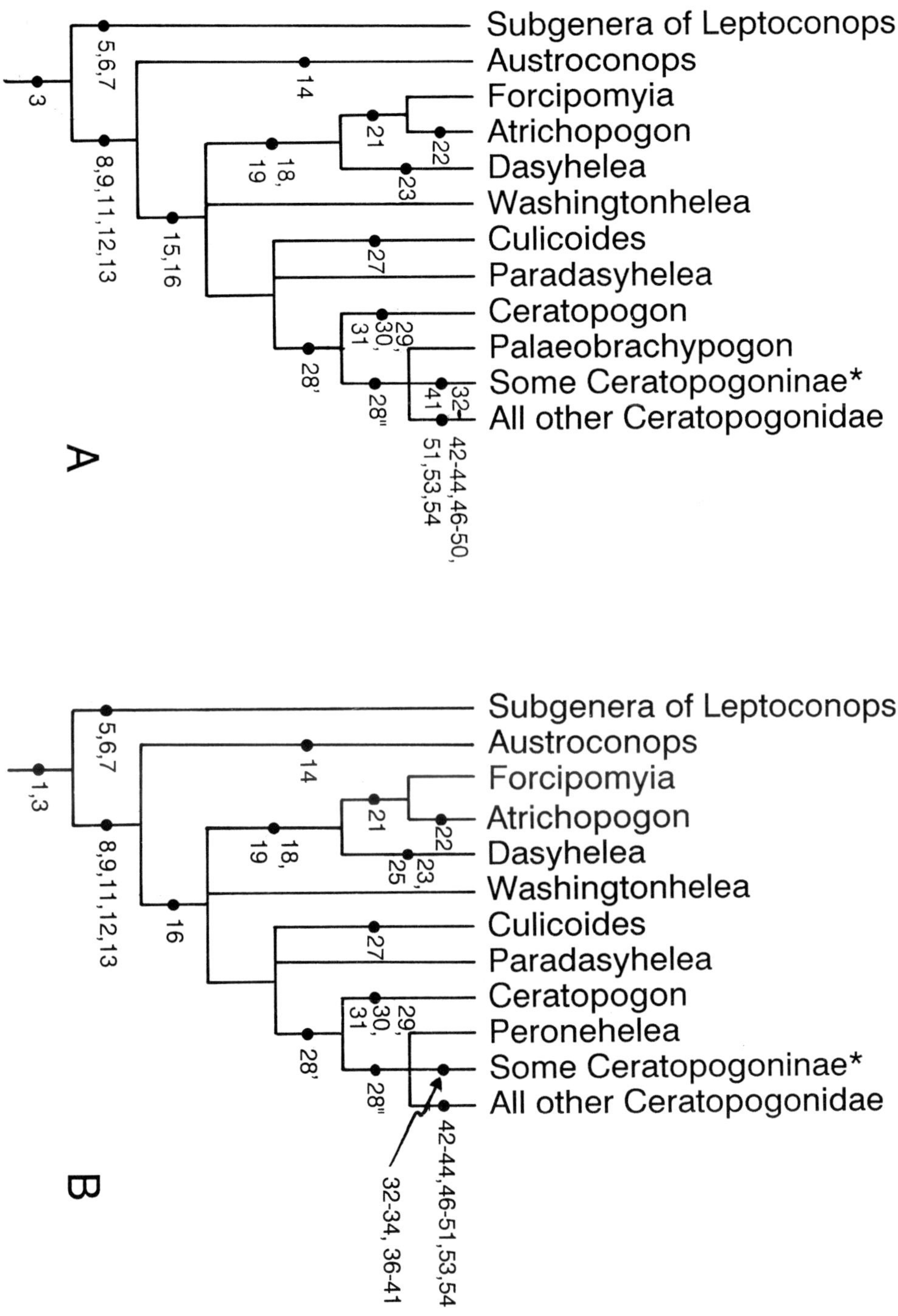
Subgenera of Leptoconops
Austroconops
Forcipomyia
Atrichopogon
Dasyhelea
Washingtonhelea
Culicoides
Paradasyhelea
Ceratopogon
Palaeobrachypogon
Some Ceratopogoninae*
All other Ceratopogonidae
A
Subgenera of Leptoconops
Austroconops
Forcipomyia
Atrichopogon
Dasyhelea
Washingtonhelea
Culicoides
Paradasyhelea
Ceratopogon
Peronehelea
Some Ceratopogoninae*
All other Ceratopogonidae
B

Fig. 25. Cladistic relationships of Upper Cretaceous and extant lineages of Ceratopogonidae. All Upper Cretaceous taxa are 71-84 million years old, except for *Austroconops* which is 93-99 million years old. Relationships reflect those supported by synapomorphies show in Fig. 21. * Refers to unresolved lineages listed in caption of Fig. 21.

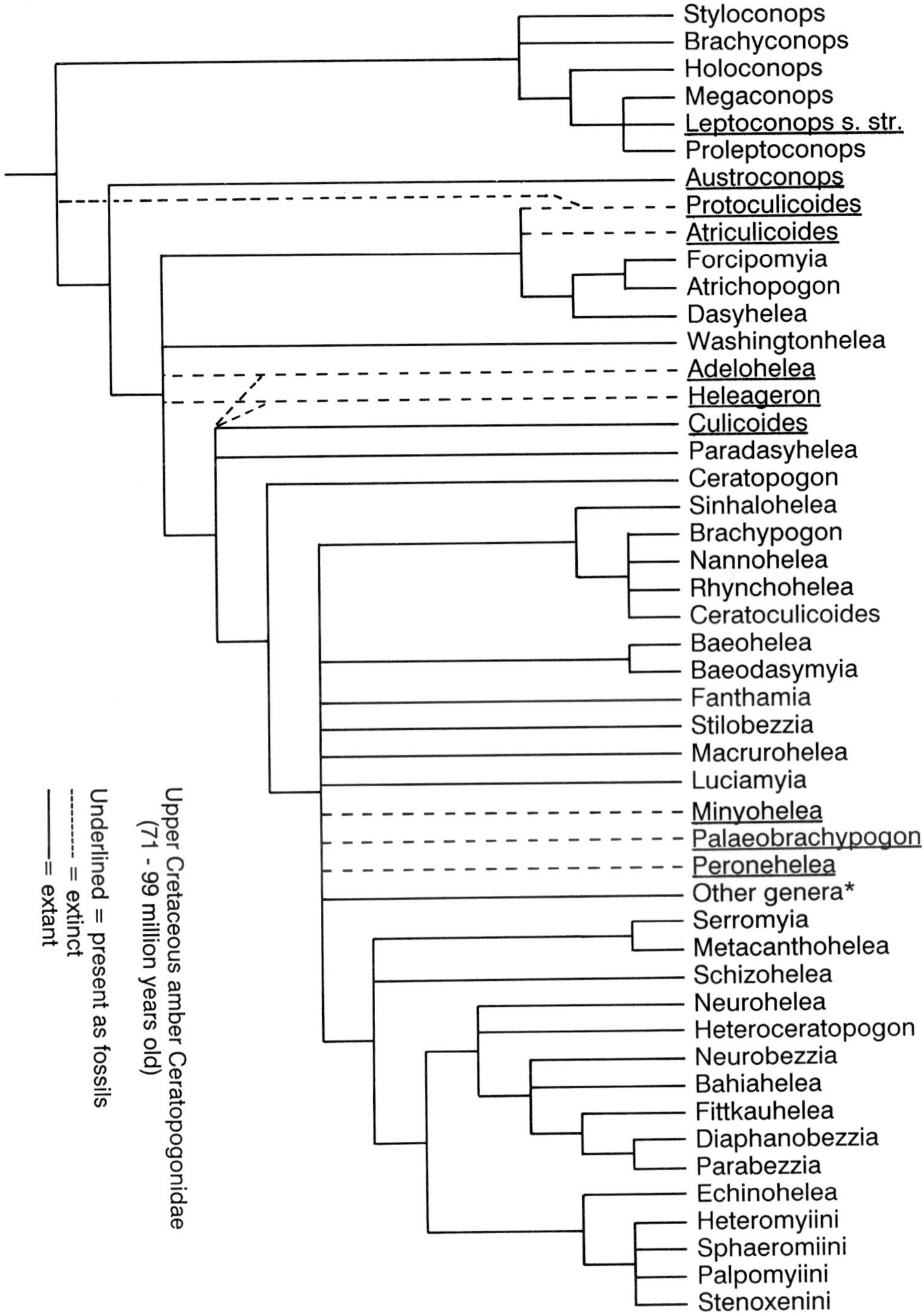
Styloconops
Brachyconops
Holoconops
Megaconops
Leptoconops s. str.
Proleptoconops
Austroconops
Protoculicoides
Atriculicoides
Forcipomyia
Atrichopogon
Dasyhelea
Washingtonhelea
Adelohelea
Heleageron
Culicoides
Paradasyhelea
Ceratopogon
Sinhalohelea
Brachypogon
Nannohelea
Rhynchohelea
Ceratoculicoides
Baeohelea
Baeodasymyia
Fanthamia
Stilobezzia
Macrurohelea
Luciamyia
Minyohelea
Palaeobrachypogon
Peronehelea
Other genera*
Serromyia
Metacanthohelea
Schizohelea
Neurohelea
Heteroceratopogon
Neurobezzia
Bahiahelea
Fittkauhelea
Diaphanobezzia
Parabezzia
Echinohelea
Heteromyiini
Sphaeromiini
Palpomyiini
Stenoxenini
Upper Cretaceous amber Ceratopogonidae
(71 - 99 million years old)
Underlined = present as fossils
-------- = extinct
——— = extant

Fig. 26. Cladistic relationships of Baltic amber and extant lineages of Ceratopogonidae. Relationships reflect those supported by synapomorphies show in Fig. 21. * Refers to unresolved lineages listed in caption of Fig. 21. ** Refers to presence of extant genera in listed tribes.

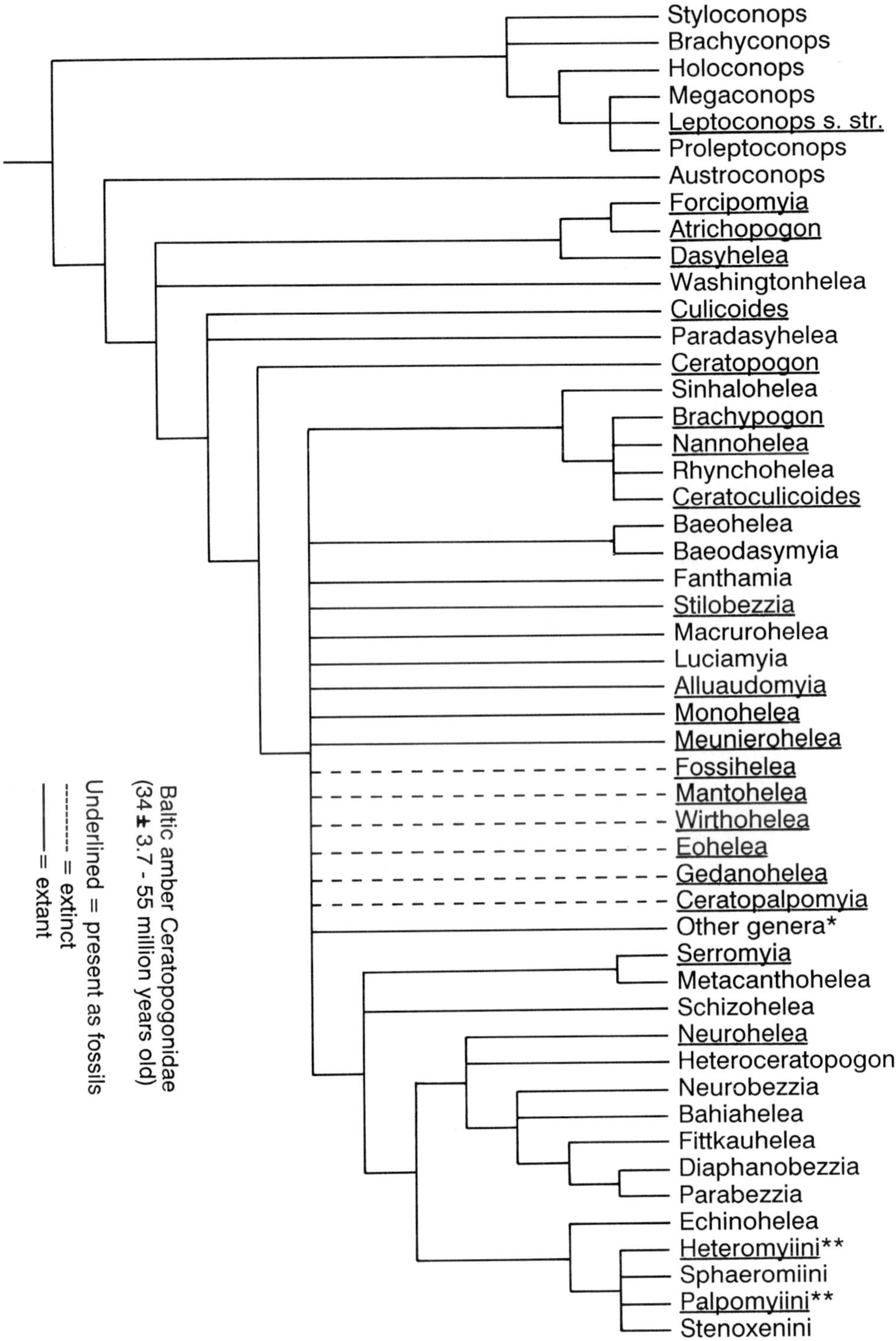
Styloconops
Brachyconops
Holoconops
Megaconops
Leptoconops s. str.
Proleptoconops
Austroconops
Forcipomyia
Atrichopogon
Dasyhelea
Washingtonhelea
Culicoides
Paradasyhelea
Ceratopogon
Sinhalohelea
Brachypogon
Nannohelea
Rhynchohelea
Ceratoculicoides
Baeohelea
Baeodasymyia
Fanthamia
Stilobezzia
Macrurohelea
Luciamyia
Alluaudomyia
Monohelea
Meunierohelea
Fossihelea
Mantohelea
Wirthohelea
Eohelea
Gedanohelea
Ceratopalpomyia
Other genera*
Serromyia
Metacanthohelea
Schizohelea
Neurohelea
Heteroceratopogon
Neurobezzia
Bahiahelea
Fittkauhelea
Diaphanobezzia
Parabezzia
Echinohelea
Heteromyiini**
Sphaeromiini
Palpomyiini**
Stenoxenini
Baltic amber Ceratopogonidae
(34 ± 3.7 - 55 million years old)
Underlined = present as fossils
--------- = extinct
——— = extant

Fig. 27. Maxillary palps of species of Nearctic *Culicoides* arranged according to host type and shape (from most slender to most distended). A, *C. furens*; B, *C. variipennis*; C, *C. hollensis* (Melander and Brues); D, *C. mississippiensis* Hoffman; E, *C. biguttatus* (Coquillett); F, *C. insignis* Lutz; G, *C. bickleyi* Wirth and Hubert; H, *C. barbosai* Wirth and Blanton; I, *C. floridensis* Beck; J, *C. baueri* Hoffman; K, *C. alexanderi* Wirth and Hubert; L, *C. obsoletus* (Meigen); M, *C. pusillus* Lutz; N, *C. tissoti* Wirth and Blanton; O, *C. snowi* Wirth and Jones; P, *C. variipennis*; Q, *C. loughnani* Edwards; R, *C. guttipennis*; S, *C. sanguisuga*; T, *C. arboricola*; U, *C. niger* Root and Hoffman; V, *C. lahillei* (Iches); W, *C. hinmani* Khalaf; X, *C. melleus* (Coquillett); Y, *C. chiopterus* (Meigen); Z, *C. paraensis* (Goeldi); AA, *C. travisi* Vargas; AB, *C. spinosus* Root and Hoffman; AC, *C. piliferus* Root and Hoffman; AD, *C. mulrennani* Beck; AE, *C. haematopotus* Malloch; AF, *C. villosipennis* Root and Hoffman; AG, *C. stellifer* (Coquillett); AH, *C. downesi* Wirth and Hubert; AI, *C. stilobezzioides* Foote and Pratt; AJ, *C. sphagnumensis* Williams; AK, *C. scanloni* Wirth and Hubert; AL, *C. ousairani* Khalaf; AM, *C. crepuscularis* Malloch; AN, *C. testudinalis* Wirth and Hubert; AO, *C. denticulatus* Wirth and Hubert; AP, *C. wisconsinensis* Jones; AQ, *C. dickei* Jones; AR, *C. utowana* Jamnback; AS, *C. juddi* Cochrane; AT, *C. flukei* Jones; AU, *C. loisae* Jamnback; AV, *C. bermudensis* Williams; AW, *C. furensoides* Williams; AX, *C. pechumani* Cochrane.

C. variipennis is shown twice, likely representing variation due to more than one species under this name. Figures G, K, L, S, AH - AJ, AO - AR, AT, AU, and AW from Jamnback (1965); all others from Blanton and Wirth (1979). Host records were mainly from Blanton and Wirth (1979) and Knausenberger (1987:569-570). Species with unknown hosts are not shown.

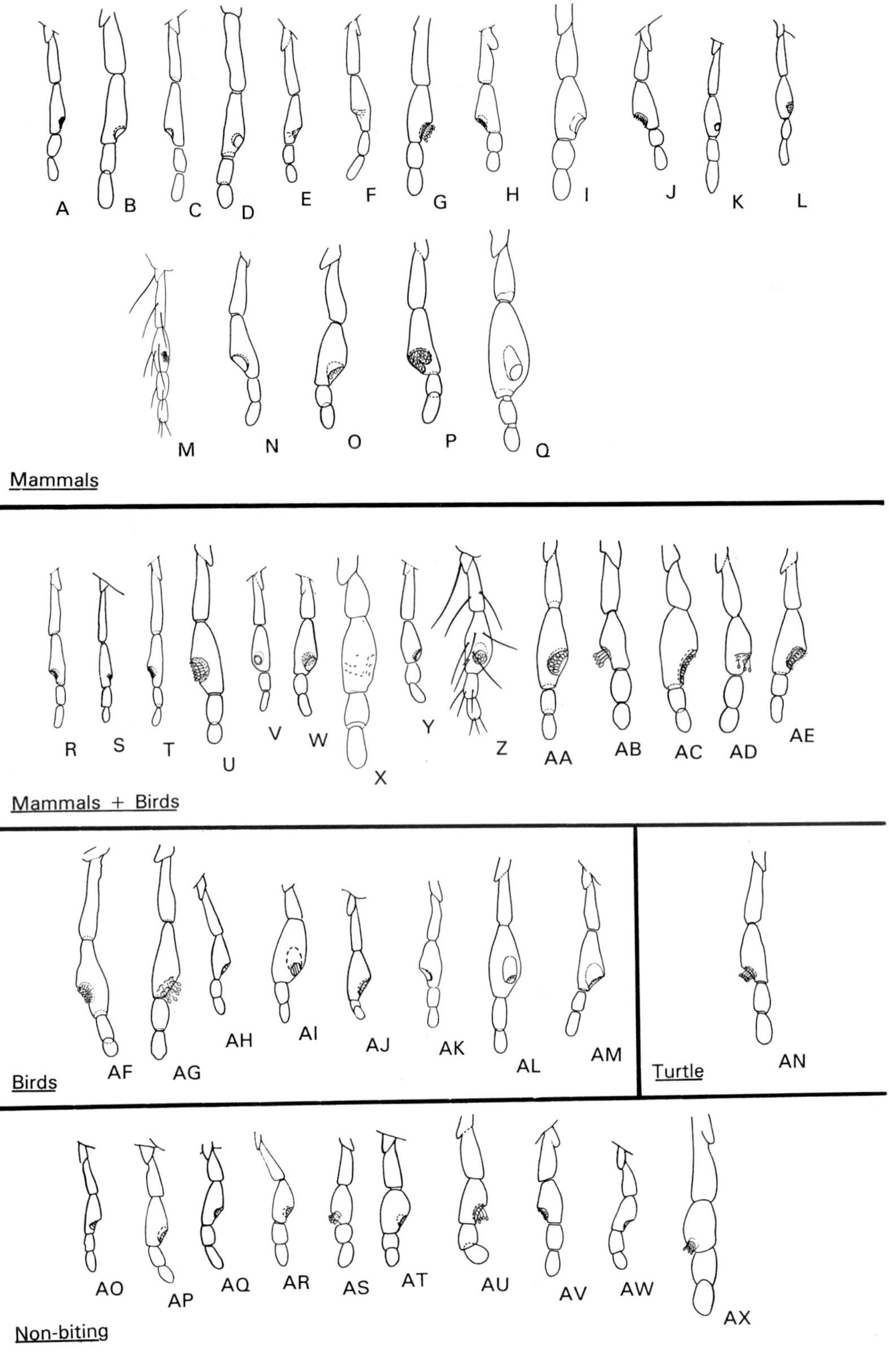
A B C D E F G H I J K L
M N O P Q
Mammals
R S T U V W X Y Z AA AB AC AD AE
Mammals + Birds
AF AG AH AI AJ AK AL AM
Birds
Turtle
AN
AO AP AQ AR AS AT AU AV AW AX
Non-biting

INDEX OF TAXONOMIC NAMES

Page numbers of principle entries are in boldface. Synonyms are in italics. Citations in keys, figures and in Table 18 (catalogue of all fossil Ceratopogonidae) are not included.